STP 1414

Pesticide Formulations and Application Systems: A New Century for Agricultural Formulations, Twenty First Volume

Jane C. Mueninghoff, Alan K. Viets, and Roger A. Downer, editors

ASTM Stock Number: STP1414

ASTM
100 Barr Harbor Drive
PO Box C700
West Conshohocken, PA, 19428-2959

Printed in the U. S. A.

ISBN: 0-8031-2891-6
ISSN: 1040-1695

Peer Review Policy

Each paper published in this volume was evaluated by two peer reviewers and at least one editor. The authors addressed all of the reviewers' comments to the satisfaction of both the technical editor(s) and the ASTM Committee on Publications.

To make technical information available as quickly as possible, the peer-reviewed papers in this publication were prepared "camera-ready" as submitted by the authors.

The quality of the papers in this publication reflects not only the obvious efforts of the authors and the technical editor(s), but also the work of the peer reviewers. In keeping with longstanding publication practices, ASTM maintains the anonymity of the peer reviewers. The ASTM Committee on Publications acknowledges with appreciation their dedication and contribution of time and effort on behalf of ASTM.

Printed in Bridgeport, NJ
November 2001

Foreword

This publication, Pesticide Formulations and Application Systems: A New Century for Agricultural Formulations, Twenty First Volume, contains selected papers presented at the symposium of the same name held in Orlando, Florida, on 24–26 October 2000. The symposium was sponsored by Committees E-35 and E-35.22 on Pesticide Formulations and Application Systems. The symposium chair was Jane C. Mueninghoff, Huntsman Corporation. The symposium co-chairs were Alan K. Viets, Bayer Corporaton and Roger A. Downer, Ohio State University.

Contents

Overview

The Twenty First Symposium on Pesticide Formulations and Applications Systems was held October 24 and 25, 2001 in Orlando, Florida. Not only was this event the twenty first symposium sponsored by the ASTM E35.22 subcommittee, it was the first symposium of the new century. Hence, the theme was chosen was "A New Century for Agricultural Formulations." The symposium program emphasized the external factors affecting the global growth of the agricultural industry as well as new options available for formulating and delivering agricultural products to meet this global growth. The twenty first symposium was a valuable forum for formulators, researchers, applicators, and marketers to share ideas, opinions, and knowledge on a variety of important topics.

Invited speakers tackled the complicated subjects of 1) the effect of biotechnology on the food industry and consumer and government response to changes in how our food supply is generated; 2) recent trends in pesticide formulations in Japan and the Asia-Pacific region; and 3) recent regulatory-actions of the United States Environmental Protection Agency. While these papers are not included in this book, extensive interest and discussion was generated by these presentations.

Papers from the technical sessions covering efficacy, methodology, formulation strategy, formulation ingredients, regulatory issues and application, and spray drift can be found in this publication.

Efficacy

This session was intended to present studies relating to efficacy enhancement of pesticide products. The presentations and papers found in this book cover the effects of adjuvants, surfactants, and inorganic cations on pesticide activity.

Methods

Techniques and methodology for determining the stability and efficacy of pesticide formulations are important tools not to be overlooked by the developer of agricultural products. This session addressed various issues of methodology, including new techniques under consideration in Europe.

Formulation Strategy

Procedures used by formulators to develop pesticide and adjuvant products are ever changing and progressing in order to decrease development time and improve optimization steps. Four papers from this session are in this STP, covering various statistical design techniques and other formulating options.

Formulation Ingredients

The development and application of inert ingredients for agricultural formulations are areas of high interest to symposium attendees. Inert ingredients can improve formulation stability, efficacy, and handling. These effects are discussed in six papers from the two sessions on formulation ingredients. Some of the inerts discussed are new to agricultural use, while others are completely new chemistry or new applications for known chemistry.

Regulatory Issues

Actions taken by governmental authorities can have broad impact not just on pesticide active ingredients, but on inert ingredients and formulation additives as well. While only one of the papers from this session is included in this book, the topic of environmental strategies for nonylphenol ethoxylates is certainly of great interest to formulators around the world.

Application and Spray Drift

Another key component of a successful agricultural product is its application in the field. This session concentrated on spray drift characteristics, testing, measurement, and distribution. Each paper in this area discusses useful measurement techniques and effects on pesticide efficacy.

Jane C. Mueninghoff
Symposium Chair
Huntsman Corporation
Houston, Texas

SESSION I: Efficacy

Hans de Ruiter,[1] Els Nijhuis,[1] Horst-Werner Wollenweber,[2] and Hans-Georg Mainx[2]

Cellular Toxicity of Different Classes of Adjuvants

Reference: de Ruiter, H., Nijhuis, E., Wollenweber, H. W., and Mainx, H. G., **"Cellular Toxicity of Different Classes of Adjuvants,"** *Pesticide Formulations and Application Systems: A New Century for Agricultural Formulations, Twenty First Volume, ASTM STP 1414,* J. C. Mueninghoff, A. K Viets, and R. A. Downer, Eds., American Society for Testing and Materials, West Conshohocken, PA, 2001.

Abstract: The phytotoxicity of adjuvants depends on the amount of surfactant deposited per unit of leaf area, the penetration of the surfactant into the leaf, and the cellular toxicity. In this study we investigated the cellular toxicity by determining the leakage of electrolytes from potato discs. We included four adjuvant classes [alkylpolyglycosides (APG), polyoxyethylene fatty alcohols (FAL-EO), methylated polyoxyethylene fatty acids (FAC-EO-ME), and esterified polyoxyethylene glycerols (G-EO-E),]. In each class we included a lipophilic and a hydrophilic product by variation of the C–content of the alkyl chain (APG) or EO-content (other classes). Conductivity of the bathing media was measured at concentrations of 10^{-4}, 10^{-3}, 10^{-2}, 10^{-1}, and 1% (w/v). A polyoxyethylene (10) nonylphenol was included as a reference compound. The cellular toxicity of the adjuvants started at concentrations >0.01% (FAL-EO and FAC-EO-ME) or >0.1% (APG).

The glycerol derivatives (G-EO-E) were not toxic at all concentrations tested. The FAC-EO-ME class of adjuvants was the most toxic class; on average 65% of the total electrolyte content of the discs was released within 3h. The FAL-EO and the APG adjuvants were less toxic excepted the C_8-C_{10} APG giving a release of 56% at a concentration of 1%. In each adjuvant class the lipophilic product was less toxic than the more hydrophilic product. The results are discussed in relation to the physical-chemical properties of the adjuvants.

Keywords: phytotoxicity, cell membrane, surfactant

The phytotoxicity of adjuvants depends on the amount of surfactant deposited per unit of leaf area, the penetration of the surfactant into the leaf, and the cellular toxicity. This damage can have deleterious effects on the translocation of pesticides in the tissue

---

[1] Senior Scientist and Research Assistant, respectively, Plant Research International, P.O. Box 16, NL-6700 AA Wageningen, The Netherlands.

[2] Manager, Product Development, and Manager, Technical Sales, respectively, Cognis Agrosolutions, Düsseldorf, Germany.

beneath the drop deposit (Sherick et al. 1986). Another aspect is that for cosmetic reasons plant growers do not want to sell products with necrotic spots on the surface (fruits, pot plants, vegetables etc.). In a previous study (de Ruiter et al. 1998) we tested a range of different surfactants with regard to their phytotoxicity on tomato (*Solanum lycopersicum* L.) and wheat (*Triticum aestivum* L.) plants. When applied at 0.5% in the treatment solution, the alkylpolyglycosides (APG) and the esterified polyalkoxy glycerols [G-EO(PO)-E] were not phytotoxic, the polyalkoxy fatty alcohols [FAL-EO(PO)] were rather phytotoxic on both species, and the methylated polyalkoxy fatty acids (FAC-EO-ME) were phytotoxic only to tomato.. After passage of the leaf cuticle, the subsequent interaction of a surfactant with the cell membrane may result in cell damage. In order to understand better the relation between surfactant structure and phytotoxicity, we investigated the cellular toxicity of a selection of surfactants from the four classes mentioned above. The cellular toxicity was characterized by determining the leakage of electrolytes from potato discs incubated in medium that contained surfactant.

Experimental Method

Plant Material and Leakage Experiment

Potato discs (*Solanum tuberosum* L. cv. Bintje) were cut (diam. 11 mm, thickness 2 mm) and washed twice with 100 mL 0.4 M mannitol. Then two discs were transferred to a 20 ml-vial with 10 mL bathing medium containing 0.4 M mannitol alone (control) or 0.4 M mannitol plus surfactant at 10^{-4}, 10^{-3}, 10^{-2}, 10^{-1}, 1 and 0% (w/v). The vials were closed and gently shaken during three hours at 25°C. Then, the conductivity (C_{180}) was measured and corrected for the conductivity of the bathing media themselves (C_0). After the incubation, the vials are placed in a water bath at 95°C during 20 minutes to disrupt the cell integrity. The conductivity is measured again after cooling to 25°C. This conductivity after disruption (C_{max}) is corrected for the conductivity of the bathing media themselves (C_0) and indicates the total electrolyte content of the potato-disc. The leakage (L) induced by the surfactants is calculated according to

$$L = [(C_{180}-C_0)/(C_{max}-C_0)] \times 100\%$$

Surfactants

By variation in ethylene oxide (EO) content or size of the alkyl chain we selected per surfactant class two products differing in lipohilicity (Table 1). A polyoxyethylene (10) nonylphenol surfactant was included as a reference compound.

Table 1- *Surfactants included in this study.*

Class	Product name[1]	C-content alkyl chain	EO-content	cmc[2] (%)
APG	Agrimul® PG 2076	8-10	-	0.02
APG	Agrimul® PG 2062	12-14	-	0.002
FAL-EO	Dehydol® LS 5	12-14	5	0.005
FAL-EO	Dehydol® LS 9.5	12-14	9.5	n.d.[3]
FAC-EO-ME	Eumulgin® ME 3516	6-10	3	0.005
FAC-EO-ME	Eumulgin® ME 3517	6-10	6	0.008
G-EO-E	Eumulgin® CO 3371	16-18	5	n.d.
G-EO-E	Eumulgin® CO 3393	16-18	20	0.7
AP-EO	Dehydrophen® 100	9	10	0.007

[1]All products were obtained from Cognis Agrosolutions, Düsseldorf, Germany.
[2]Critical micelle values (cmc) were provided by the producer.
[3]n.d.= not determined

Experimental Design

Three separate experiments with each two replicates were conducted. Variables in each replicate were nine surfactant types and six concentrations per surfactant type. Discs from two potatoes were used for one replicate and were washed as described above. For each separate experiment we prepared new surfactant solutions. The sequence of treatments at the start of the incubations was randomized. After angular transformation the data were subjected to analysis of variance using the Genstat[3] statistical package.

Results and Discussion

In preparatory experiments with Dehydol® LS 9.5 and Agrimul® PG 2076, we measured an almost linear increase of electrolyte leakage when the leakage was determined after different incubation periods up to 4.5 h (data not shown). Therefore, we decided to select an incubation period of 3 h as a standard incubation period for all further testing. Mannitol was included in all our treatment solutions because this lowered the disruption of cells as a result of the experimental procedure itself as was demonstrated in preparatory experiments.

[3] Genstat 5 release 3.2 from Lawes Agricultural Trust. Rothamsted Experimental Station, Harpenden, Herts AL5 2 JQ, England.

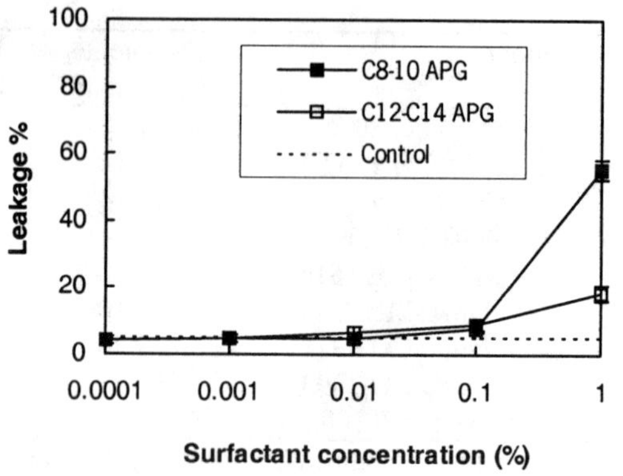

Figure 1-*Influence of Agrimul® PG 2076 ($C_{8\text{-}10}$ APG) and Agrimul PG 2062 ($C_{12\text{-}14}$ APG) on the leakage of electrolytes from potato-discs.*

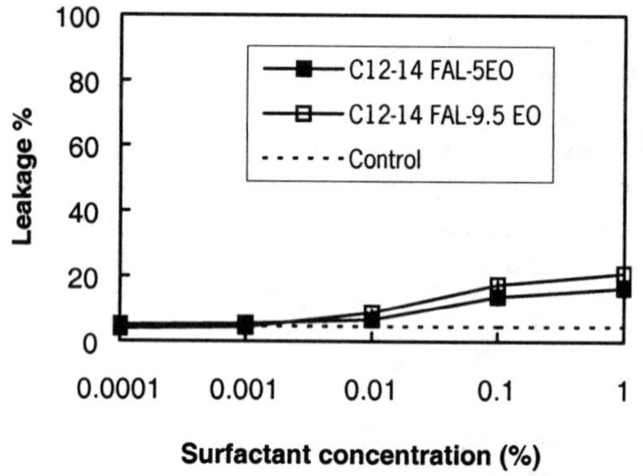

Figure 2-*Influence of Dehydol® LS 5($C_{12\text{-}14}$ FAL-5EO) and Dehydol LS 9.5 ($C_{12\text{-}14}$ FAL-9.5EO) on the leakage of electrolytes from potato-discs.*

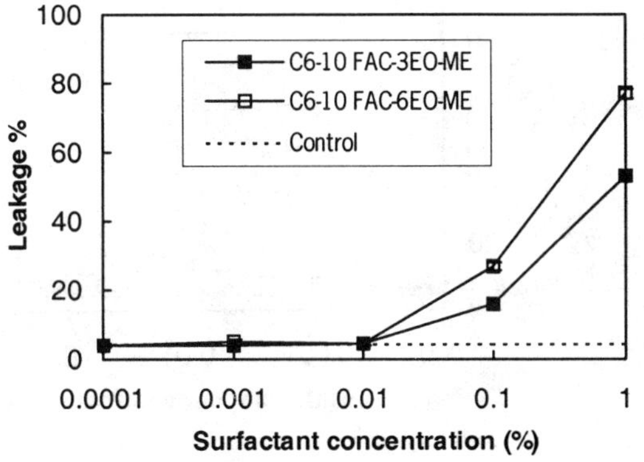

Figure 3-*Influence of Eumulgin® ME 3516 (C_{6-10} FAC-3EO-ME) and Eumulgin ME 3517 (C_{6-10} FAC-6EO-ME) on the leakage of electrolytes from potato-discs.*

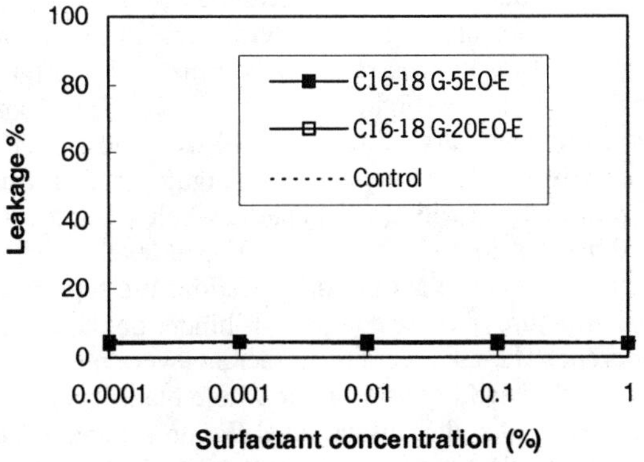

Figure 4-*Influence of Eumulgin® CO 3371 (C_{16-18} G-5EO-E) and Eumulgin CO 3393 (C_{16-18} G-20EO-E) on the leakage of electrolytes from potato-discs.*

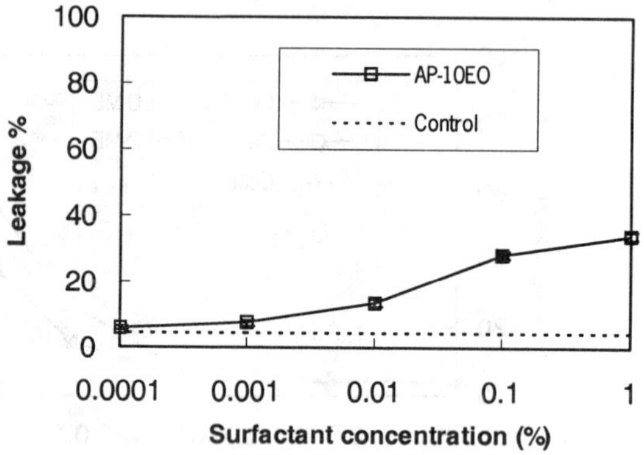

Figure 5-*Influence of Dehydrophen® 100 (AP-10EO) on the leakage of electrolytes from potato-discs.*

The critical micelle concentrations (Table 1) do not predict the concentration where the surfactants start to induce leakage of electrolytes. Alone for the fatty alcohol (Figure 2) and the fatty acid surfactants (Figure 3), there seems to be a correlation.

The APG surfactants, at concentrations lower than 0.1% (Figure 1), and the glycerol derivatives at all concentrations (Figure 4) were not toxic to the potato discs. The methylated polyoxyethylene fatty acids (Figure 3) were most toxic of all adjuvants tested, followed by the polyoxyethylene fatty alcohols (Figure 2). The results with the APG and the glycerol derivatives agree with the data at the whole plant level. Alone APG at a concentration of 0.5%was not toxic at the whole plant level whereas the cellular toxicity data indicate membrane damage at this concentration. We suggest that the molecular three-dimensional structure of these compounds hinder passage over both the leaf cuticle and the cell membrane. The fatty alcohol surfactants were much more toxic than the methylated polyoxyethylene fatty acids at the whole plant level (de Ruiter et al. 1998). This outcome was reversed at the cellular level (Figures 2 and 3; Table 2). The difference between results at the whole plant level and at the cellular level suggests that the end-capping of the EO-chain and/or the additional carbonyl had such an influence on the physical-chemical properties of the fatty acid derivatives that these compounds are retained on or in the leaf cuticle. The high level of phytotoxicity of the fatty alcohol derivatives at the whole plant level and their relatively low level of cellular toxicity indicates that these compounds easily pass the leaf cuticle. The suggested low foliar penetration of the fatty acid derivatives may explain their low level of phytotoxicity at the whole plant level. In this reasoning, we assumed that the cellular toxicity as observed with the potato disc are also valid for other species. The reference compound polyoxyethylene (10) nonylphenol was toxic to the potato-discs (Figfure 5). This compound was also toxic after application of this compound to whole plants (de Ruiter et al. 1998) and agrees with previous data on the polyoxyethylene nonylphenol type surfactants as reviewed by Gaskin (1995).

In each pair of the adjuvant classes the more lipophilic surfactant was less toxic than the more hydrophilic compound (Figures 1, 2, 3, and 4; Table 1). At this stage we suggest that the more lipophilic surfactants form micelles and other aggregates at a lower concentration than the more hydrophilic surfactants. As a consequence a higher concentration of monomers may interact with cell membranes in the case of the more hydrophilic surfactants.

Table 2-*The influence of the surfactants on the leakage of electrolytes.*

		Leakage after 3 h incubation[1], %		
		Surfactant concentration, %		
Class	Product name	0.01	0.1	1
APG	Agrimul® PG 2076	5	8 a	56 h
APG	Agrimul® PG 2062	7 a[3]	9 a	18 de
FAL-EO	Dehydol® LS 5	7 a	14 bc	16 cd
FAL-EO	Dehydol® LS 9.5	8 a	17 d	21 e
FAC-EO-ME	Eumulgin® ME 3516	5	16 bcd	53 h
FAC-EO-ME	Eumulgin® ME 3517	4	27 f	77 i
G-EO-E	Eumulgin® CO 3371	4	5	4
G-EO-E	Eumulgin® CO 3393	4	4	4
AP-EO	Dehydrophen® 100	13 b	29 f	34 g
	Control[2]	4	4	4

[1]See section on Experimental Methods for definition of leakage.
[2]Control is leakage after incubation without surfactant.
[3]All means marked with letters differ from the control at the 5% probability level. Means followed by the same letter do not differ significantly.

A comparison of all compounds at 0.01, 0.1% and 1% (Table 2) demonstrates the absence of toxicity of the glycerol derivatives at all concentrations, and the increase of toxicity of the APG class, particularly the C_{8-10} variant, when increasing their concentration up to 1%. Particularly at 1%, the fatty acid derivatives are more toxic than the fatty alcohol derivatives whereby the 6-EO fatty acid derivative was much more toxic than the 3-EO variant.

Our study demonstrates that cellular toxicity data can, but do not necessarily agree with the whole plant toxicity data. We provide indirect evidence that the structure of a surfactant may have a great influence on its foliar penetration and cellular toxicity and thus on the possibility of leaf necrosis. We suggest that variations in the lipophilicity, the molar volume, and the 3-dimensional structure of the molecule are worthwhile parameters to investigate further for optimization of their application as adjuvants for agrochemicals.

References

De Ruiter, H. Mainx, H. G., Wollenweber, H. W., Uffing, A. J. M., and Withagen, J. C. M., 1998, "Phytotoxicity of Different Classes of Adjuvants," *Proceedings of 5th International Symposium on Adjuvants for Agrochemicals*, P. McMullan, ed., Vol. 1, pp. 204-209.

Gaskin, R., 1995, "Phytotoxicity of Agrochemical Surfactants", *Proceedings of 4th International Symposium on Adjuvants for Agrochemicals*, R. Gaskin, ed., Forest Research, Rotura, New Zealand, pp. 193-200.

Sherrick, S. L., Holt, H. A., and Hess, F. D., 1986, " Effects of Adjuvants and Environment during Plant Development on Glyphosate Absorption and Translocation in Field Bindweed (*Convolvulus arvensis*). *Weed Science*, Vol. 34, pp. 811-816.

Zenon Woznica,[1] John D. Nalewaja,[2] and Calvin G. Messersmith[2]

Sulfosulfuron Efficacy Is Affected by Surfactants, pH of Spray Mixture, and Salts

---

Reference: Woznica, Z., Nalewaja, J. D., and Messersmith, C. G., **"Sulfosulfuron Efficacy Is Affected by Surfactants, pH of Spray Mixture, and Salts,"** *Pesticide Formulations and Application Systems: A New Century for Agricultural Formulations, Twenty First Volume, ASTM STP 1414,* J. C. Mueninghoff, A. K. Viets, and R. A. Downer, Eds., American Society for Testing and Materials, West Conshohocken, PA, 2001.

Abstract: Sulfosulfuron {1-(4,6-dimethoxypyrimidin-2-yl)-3-[(2-ethanesulfonyl-imidazo[1,2-a]pyridine-3-yl)]sulfonylurea} must be applied with an adjuvant for effective postemergent weed control. Experiments were conducted to determine sulfosulfuron phytotoxicity to green foxtail (*Setaria viridis* L.) as influenced by surfactant chemistry, surfactant hydrophilic/lipophilic balance (HLB), spray mixture pH, and common nitrogen fertilizers. Regardless of accompanying surfactant chemistry, surfactant HLB, and nitrogen fertilizers, sulfosulfuron efficacy to green foxtail was greatly enhanced when applied in alkaline spray mixture (pH 7.4-8.2) compared to pH 4.0-5.7 unbuffered spray mixture. Secondary alcohol ethoxylate surfactants increased sulfosulfuron efficacy as HLB of these surfactants increased from 8.3 to 17.4 in unbuffered and buffered spray mixtures. The optimum HLB of various block copolymer surfactants for sulfosulfuron was from 9.0 to 13.0. In unbuffered spray mixture, ammonium nitrate at 0.25% w/v had no effect or antagonized sulfosulfuron phytotoxicity, especially when applied with block copolymer surfactants. However, urea fertilizer at 0.25% w/v enhanced herbicide efficacy, regardless of surfactants. Commercial 28% N liquid fertilizer (urea/ammonium nitrate solution) also increased sulfosulfuron efficacy with most surfactants, especially in unbuffered spray mixtures. In spray mixtures buffered above pH 7, the effects of surfactant HLB and nitrogen fertilizers on sulfosulfuron efficacy were less evident. Spray droplet spread and droplet drying time did not relate to sulfosulfuron efficacy.

These data demonstrate a potential for optimizing sulfosulfuron efficacy by carefully selecting surfactant chemistry, surfactant HLB, addition of proper nitrogen fertilizer, and by regulating the pH of the spray mixture.

Keywords: adjuvant, HLB, ammonium nitrate, urea, 28% N liquid fertilizer, triethanolamine, spray pH, green foxtail

---

[1]Visiting Scientist, North Dakota State University, Dept. of Plant Sciences, Fargo, ND, 58105. Permanent address: Professor, Agricultural University, Dept. of Agronomy, 60-623 Poznan, Poland.

[2]Professors, North Dakota State University, Dept. of Plant Sciences, Fargo, ND, 58105.

Adjuvants added to herbicide spray mixtures applied postemergence often increase spray retention and/or foliar absorption, which results in enhanced herbicide activity (McWhorter 1982). Adjuvant impacts on efficacy of postemergence herbicides differs greatly depending on the herbicide solubility, adjuvant physical and chemical properties, hydrophilic/lipophilic balance (HLB), adjuvant concentration, salts present in spray water carrier, and spray mixture pH (Green 1999b, Manthey et al. 1998, Nalewaja et al. 2000).

Nitrogen-containing fertilizers, particularly ammonium sulfate and liquid urea/ammonium nitrate fertilizer with 28% nitrogen, have been widely used as adjuvants for directly enhancing activity of many postemergence herbicides (Gronwald et al. 1993) or to overcome herbicide antagonism by salts in the water carrier (Nalewaja and Matysiak 1993, Nalewaja et al. 1998). However, enhancement of nicosulfuron efficacy by 28% nitrogen fertilizer has been inconsistent and dependent on the accompanying surfactant or oil and plant species tested (Lueschen et al. 1991, Morton et al. 1991). Nalewaja et al. (1998) found that urea never enhanced and often reduced surfactant and oil adjuvant enhancement of nicosulfuron. However, ammonium nitrate enhanced nicosulfuron phytotoxicity to yellow foxtail [*Setaria glauca* (L.) P. Beauv.], especially when the spray carrier contained sodium bicarbonate. Based on this research, the authors concluded that the benefit to nicosulfuron from commercial 28% liquid nitrogen fertilizer was mainly from the ammonium nitrate component.

Over 100 formulations of liquid nitrogen fertilizers exist on the U.S. market, which differ in composition and ammonia ion content (Joost 1998). Most of these fertilizers increase the pH of the spray mixture, because the formulations contain various components added to minimize metal corrosion. Nalewaja et al. (2000) found that nicosulfuron efficacy was greatly increased when pH of the spray mixture was increased from 4 to 9. The efficacy increase theoretically was due to higher water solubility of nicosulfuron at high pH (39 200 mg/L at pH 8.8) than low pH spray (360 mg/L at pH 5) (Herbicide Handbook 1994).

Sulfosulfuron is a new sulfonylurea herbicide introduced for control of annual and perennial grasses and some broadleaf weeds in wheat (*Triticum aestivum* L.) (Monsanto Company, St. Louis, MO). Enhancement of sulfosulfuron efficacy by surfactants differed greatly depending on surfactant chemistry and HLB, inclusion of ammonium nitrate, and weed species (Woznica et al. 1997). Octylphenol, nonylphenol, and secondary alcohol ethoxylate surfactants generally enhanced sulfosulfuron efficacy to green foxtail as HLB value of surfactants increased. However, the ammonium nitrate effect on sulfosulfuron efficacy differed greatly depending upon the surfactant and weed species. Miller et al. (1999) found that 28% nitrogen liquid fertilizer greatly increased absorption of sulfosulfuron by most weed species. Sulfosulfuron solubility in water, as with other sulfonylurea herbicides, varies with the spray mixture pH (18 mg/L at pH 5, 1627 mg/L at pH 7, and 482 mg/L at pH 9) (Herbicide Handbook 1998). Thus, a proper spray mixture pH in combination with surfactant and/or nitrogen fertilizer may be important to sulfosulfuron efficacy.

Experiments were conducted to determine sulfosulfuron efficacy to green foxtail as influenced by ammonium nitrate, urea, and laboratory-made and commercial 28% nitrogen ammonium nitrate/urea liquid fertilizer applied at low and increased pH of spray mixture.

Experimental Method

Green foxtail was seeded in 4-cm diameter by 20-cm long plastic cones containing a commercial peat-based greenhouse mixture. Plants were thinned to seven per cone 1 wk after emergence and were watered and fertilized as needed for healthy growth. Natural daylight was supplemented for 16 h photoperiod with metal halide lamps with the light intensity of 450 $\mu E/m^2/s$. The greenhouse was maintained at 20 °C at night and 25 ± 5 °C during the day. Sulfosulfuron at 15 g ai/ha was applied to three leaf plants using a cabinet laboratory sprayer that delivered 80 L/ha through a 8001 flat fan nozzle at 280 kPa generated by compressed air. Distilled water was the spray carrier. Herbicide uptake from the soil was prevented by covering the soil surface with vermiculite before treatment and removing it after treatment.

Surfactants tested were secondary alcohol ethoxylates with a HLB range from 8.3 to 18.0 (Tergitol® 15-S series)[3] and block copoplymers with a HLB range from 4.0 to 15.0 (Pluronic® series)[4]. All surfactants were applied at 0.25% v/v for liquid or w/v for solid products. Nitrogen was added to the spray mixtures as ammonium nitrate at 0.25% (w/v), urea at 0.25% (w/v), ammonium nitrate/urea as a laboratory-made mixture at 0.25/0.25% (w/v), and commercial fertilizer (28% N) at 1% (v/v). Triethanolamine (TEA), a basic pH buffer, was added to the spray mixture at 0 and 0.1% (v/v), except in the concentration experiment where it was added at 0, 0.02, 0.04, 0.08, 0.16, and 0.32% (v/v).

The spray mixture pH was determined at 20 °C before spraying. Shoot fresh weight was determined 14 d after treatment, and efficacy data are presented as percentage fresh weight reduction compared to untreated plants.

Droplet spread and droplet drying time measurements were made using sulfosulfuron at 15 g ai/ha in 80 L/ha spray volume, with and without surfactants, nitrogen fertilizers, and TEA at concentrations described above. Measurements were determined using the third leaf of four-leaf green foxtail. Leaves were excised and held horizontally on the laboratory bench. A 2 μL droplet of spray mixture was deposited on the adaxial surface using a micro-syringe. Droplet spread was estimated 1 min after application based on a 1 to 5 scale, where 1 indicates low droplet spread similar to distilled water and 5 indicates complete droplet spreading. Droplet spread and droplet drying time were measured at 20 °C and 50 ± 5% relative humidity.

Experiments were conducted using a randomized complete block design with four replicates and each experiment was repeated. Means were separated using Fisher's protected LSD at 95% probability.

Results and Discussion

Distilled Water Carrier Tergitol® 15-S surfactant efficacy for enhancement of sulfosulfuron efficacy generally increased as HLB increased to 17.4 and then occasionally decreased with an HLB of 18.0 (Table 1). The lowest HLB Tergitol®

---

[3]Tergitol® 15-S surfactants are from Union Carbide, Danberry, CT.
[4] Pluronic® surfactants are from BASF Performance Chemicals, Mount Olive, NJ.

15-S-3 alone or with nitrogen salts did not increase sulfosulfuron efficacy compared to sulfosulfuron applied in distilled water alone. The 28% nitrogen fertilizer greatly increased sulfosulfuron efficacy when applied with Tergitol® 15-S surfactants having HLB values from 12.4 to 18.0 and eliminated the differences in efficacy among these surfactants. However, the laboratory-made ammonium nitrate and urea mixture prepared to simulate 28% nitrogen fertilizer, did not change sulfosulfuron efficacy with any Tergitol® 15-S surfactants from that without nitrogen salts. Thus, the formulants in 28% nitrogen fertilizer were more important than the ammonium nitrate and urea for enhancing sulfosulfuron efficacy. Sulfosulfuron was enhanced more frequently by Tergitol® 15-S surfactants when with urea than ammonium nitrate or ammonium nitrate plus urea, but less than with 28% nitrogen.

Sulfosulfuron efficacy for green foxtail control was much greater when applied with Pluronic® than Tergitol® 15-S surfactants (Tables 1 and 2). Efficacy of Pluronic® surfactants generally increased as HLB increased from 4 to 9 or 13 depending on the accompanying salt and then decreased greatly for HLB 14 and 15. Ammonium nitrate,

Table 1 – *Green foxtail fresh weight reduction from sulfosulfuron at 15 g ai/ha as influenced by secondary alcohol ethoxylate (Tergitol®15-S) surfactants at 0.25% (v/v), nitrogen fertilizers, and triethanolamine at 0.1% (v/v).*

		Nitrogen fertilizer[2]				
Treatment	HLB[1]	None	AN	U	AN/U	28% N
Distilled water carrier		% fresh weight reduction				
No surfactant		9	10	18	12	16
Tergitol® 15-S-3	8.3	14	11	17	14	17
Tergitol® 15-S-7	12.4	19	17	21	19	49
Tergitol® 15-S-9	13.3	20	22	26	22	59
Tergitol® 15-S-15	15.4	25	25	29	23	50
Tergitol® 15-S-20	16.3	28	27	43	30	53
Tergitol® 15-S-30	17.4	34	28	44	33	52
Tergitol® 15-S-40	18.0	35	33	30	24	48
Distilled water + triethanolamine carrier						
No surfactant		30	25	37	42	49
Tergitol® 15-S-3	8.3	42	53	52	59	66
Tergitol® 15-S-7	12.4	51	70	64	74	68
Tergitol® 15-S-9	13.3	64	85	78	75	79
Tergitol® 15-S-15	15.4	77	89	87	83	82
Tergitol® 15-S-20	16.3	83	90	88	90	91
Tergitol® 15-S-30	17.4	87	87	88	92	91
Tergitol® 15-S-40	18.0	85	86	82	87	87
LSD (5%)				8		

[1] HLB is hydrophilic/lipophilic balance value.
[2] AN (ammonium nitrate) at 0,25% (w/v), U (urea) at 0.25% (w/v), AN/U (ammonium nitrate/urea laboratory-made) at 0.25/0.25% (w/v), 28% N (commercial liquid nitrogen fertilizer) at 1% (v/v).

which did not influence Tergitol® 15-S surfactant effectiveness with sulfosulfuron, reduced efficacy when applied with Pluronic® surfactants and tended to reduce the optimum HLB value for efficacy. The 28% N fertilizer enhancement of Pluronic® surfactant efficacy did not change the optimum HLB for sulfosulfuron efficacy. Urea and laboratory-made ammonium nitrate/urea mixture were intermediate to 28% nitrogen fertilizer for enhancement of sulfosulfuron with all Pluronic® surfactants.

These results indicate that inclusion of nitrogen salts could increase or decrease sulfosulfuron efficacy depending on the specific salt and surfactant. The greater enhancement of sulfosulfuron efficacy from 28% N fertilizer than from any other nitrogen salt probably reflects the higher pH of the spray mixture, important for sulfosulfuron solubility (Tables 3 and 4).

Commercial 28% N is formulated with a high pH to reduce corrosion of metal. A pH change was not the reason for enhancement of sulfosulfuron efficacy from urea with intermediate HLB Tergitol® 15-S and Pluronic® surfactants or the reduced efficacy from ammonium nitrate with Pluronic® surfactants. These salts with sulfosulfuron did not change the pH of the spray mixture compared to sulfosulfuron without salts.

The differences in sulfosulfuron efficacy resulting from changes in surfactant HLB or

Table 2 – *Green foxtail fresh weight reduction from sulfosulfuron at 15 g ai/ha as influenced by block copolymer (Pluronic®) surfactants at 0.25% (v/v), nitrogen fertilizers, and triethanolamine at 0.1% (v/v).*

Treatment	HLB[1]	Nitrogen fertilizer[2]				
		None	AN	U	AN/U	28% N
Distilled water carrier		% fresh weight reduction				
No surfactant		0	0	7	6	16
Pluronic® L 122	4.0	36	29	47	46	58
Pluronic® L 62	7.0	51	42	63	60	76
Pluronic® P 103	9.0	58	41	73	60	79
Pluronic® P 104	13.0	63	29	69	66	82
Pluronic® P 84	14.0	19	7	39	36	56
Pluronic® L 64	15.0	23	6	38	32	52
Distilled water + triethanolamine carrier						
No surfactant		38	41	50	57	56
Pluronic® L 122	4.0	64	84	85	87	84
Pluronic® L 62	7.0	74	82	89	87	82
Pluronic® P 103	9.0	79	87	91	90	85
Pluronic® P 104	13.0	87	87	91	89	88
Pluronic® P 84	14.0	80	88	90	86	87
Pluronic® L 64	15.0	78	88	91	88	86
LSD (5%)				14		

[1] HLB is hydrophilic/lipophilic balance value.

[2] AN (ammonium nitrate) at 0.25% (w/v), U (urea) at 0.25% (w/v), AN/U (ammonium nitrate/urea laboratory-made) at 0.25/0.25% (w/v), 28% N (commercial liquid nitrogen fertilizer) at 1% (v/v).

nitrogen salt did not relate completely to droplet spread or drying time. Nitrogen salts did not influence droplet spread or droplet drying time with either Tergitol® 15-S or Pluronic® surfactants. Droplet spread generally decreased from about 4.1 to 1.2 (on a scale of 1 - no spread to 5 - complete spread) as Tergitol® 15-S HLB value increased from 8.3 to 18.0 (Table 5). However, the spread of droplets with Pluronic® surfactants was about 1.5 and was the same for all Pluronic® surfactants, except for Pluronic® L 122 which had a spread 2.5 (data not presented). Surfactants that did not cause droplet spreading usually were associated with efficacy, but not all non-spreading surfactant droplets were efficacious.

Drying time decreased as droplet spread increased regardless of salt or specific surfactant (data not presented). Thus, the longer moist period that was associated with non-spreading droplets again occurred with the efficacious surfactants. However, many non-effective surfactants or salt combinations produced non-spreading droplets that remained moist. These results indicate that both moisture and the proper surfactant chemical characteristics probably are needed for sulfosulfuron absorption and efficacy.

Table 3 – pH of *sulfosulfuron (15 g ai/ha) spray mixture as influenced by secondary alcohol ethoxylate (Tergitol®15-S) surfactants at 0.25% (v/v), nitrogen fertilizers, and triethanolamine at 0.1% (v/v).*

Treatment	HLB[1]	Nitrogen fertilizer[2]				
		None	AN	U	AN/U	28% N
Distilled water carrier		pH of spray mixture				
No surfactant		5.0	5.4	5.1	5.3	6.0
Tergitol® 15-S-3	8.3	4.4	5.2	4.6	5.2	5.8
Tergitol® 15-S-7	12.4	4.0	4.8	4.3	5.0	5.6
Tergitol® 15-S-9	13.3	4.4	5.2	4.4	5.2	5.7
Tergitol® 15-S-15	15.4	4.2	5.3	4.4	5.2	5.8
Tergitol® 15-S-20	16.3	4.2	5.3	4.3	5.2	5.9
Tergitol® 15-S-30	17.4	4.2	5.2	4.4	5.1	5.7
Tergitol® 15-S-40	18.0	4.6	5.4	4.5	5.2	5.7
Distilled water + triethanolamine carrier						
No surfactant		7.7	7.5	7.8	7.5	7.6
Tergitol® 15-S-3	8.3	7.5	7.6	7.6	7.6	7.8
Tergitol® 15-S-7	12.4	7.4	7.7	7.4	7.6	7.7
Tergitol® 15-S-9	13.3	7.4	7.7	7.4	7.7	7.8
Tergitol® 15-S-15	15.4	7.4	7.7	7.5	7.6	7.8
Tergitol® 15-S-20	16.3	7.4	7.7	7.5	7.8	7.8
Tergitol® 15-S-30	17.4	7.3	7.8	7.4	7.6	7.8
Tergitol® 15-S-40	18.0	7.4	7.8	7.4	7.7	7.8

[1] HLB is hydrophilic/lipophilic balance value.
[2] AN (ammonium nitrate) at 0.25% (w/v), U (urea) at 0.25% (w/v), AN/U (ammonium nitrate/urea laboratory-made) at 0.25/0.25% (w/v), 28% N (commercial liquid nitrogen fertilizer) at 1% (v/v).

Table 4 – *pH of sulfosulfuron (15 g ai/ha)* spray mixture *as influenced by block copolymer (Pluronic®) surfactants at 0.25% (v/v), nitrogen fertilizers, and triethanolamine at 0.1% (v/v).*

Treatment	HLB[1]	Nitrogen fertilizer[2]				
		None	AN	U	AN/U	28% N
Distilled water carrier		pH of spray mixture				
No surfactant		5.7	5.6	5.8	5.6	6.9
Pluronic® L 122	4.0	5.6	5.6	5.6	5.6	6.9
Pluronic® L 62	7.0	5.6	5.8	5.6	5.7	6.9
Pluronic® P 103	9.0	5.6	5.8	5.6	5.7	7.0
Pluronic® P 104	13.0	5.6	5.8	5.5	5.7	7.0
Pluronic® P 84	14.0	5.5	5.8	5.5	5.6	7.0
Pluronic® L 64	15.0	5.7	5.8	5.4	5.7	6.9
Distilled water + triethanolamine carrier						
No surfactant		7.6	7.8	8.3	7.8	7.8
Pluronic® L 122	4.0	8.0	7.9	8.2	7.9	8.0
Pluronic® L 62	7.0	8.0	8.0	8.1	8.0	8.0
Pluronic® P 103	9.0	8.2	8.0	7.6	8.0	8.0
Pluronic® P 104	13.0	8.1	8.0	8.0	8.0	8.0
Pluronic® P 84	14.0	8.0	8.0	8.0	8.0	8.0
Pluronic® L 64	15.0	8.0	8.0	8.0	8.0	7.9

[1] HLB is hydrophilic/lipophilic balance value.
[2] AN (ammonium nitrate) at 0.25% (w/v), U (urea) at 0.25% (w/v), AN/U (ammonium nitrate/urea laboratory-made) at 0.25/0.25% (w/v), 28% N (commercial liquid nitrogen fertilizer) at 1% (v/v).

High pH Spray Carrier Sulfosulfuron efficacy increased greatly when the spray mixture pH was increased by TEA, regardless of surfactant or nitrogen salt (Tables 1 and 2). Sulfosulfuron applied with Tergitol® 15-S surfactants without TEA only gave about 20 to 40% plant fresh weight reduction but with TEA increased to 40 to 85%. TEA also enhanced and nearly eliminated differences in efficacy among Pluronic® surfactants. This was especially true for Pluronic® P 84 and L 64, where sulfosulfuron with these surfactants and low pH only gave about 20% plant fresh weight reduction versus about 90% with high pH.

Efficacy of sulfosulfuron with TEA was similar or slightly enhanced by nitrogen fertilizers when with most Tergitol® 15-S or Pluronic® surfactants versus sulfosulfuron with surfactant and TEA alone (Tables 1 and 2). Tergitol® 15-S surfactants having a HLB from 16.3 to 18.0 and Pluronic® P 104 with HLB 13 were not enhanced by including nitrogen fertilizers. However, efficacy of sulfosulfuron with Pluronic® surfactants was high and may have prevented expression of any benefit from the salts. Thus, high pH from TEA reduced the need for nitrogen fertilizers with the efficacious Pluronic® surfactants. The use of nitrogen fertilizers and TEA in commercial practice would help assure sulfosulfuron efficacy since surfactant efficacy may not be known and nitrogen fertilizers never reduced efficacy of high pH mixtures.

Table 5 – Spread of 2 μL droplets of sulfosulfuron (15 g ai/ha) spray mixture on a green foxtail leaf as influenced by secondary alcohol ethoxylate (Tergitol® 15-S) surfactants at 0.25% (v/v), nitrogen fertilizers, and triethanolamine at 0.1% (v/v).

		Nitrogen fertilizer[2]				
Treatment	HLB[1]	None	AN	U	AN/U	28% N
Distilled water carrier		Spread (1-5 scale[3])				
No surfactant		1.0	1.0	1.0	1.0	1.0
Tergitol® 15-S-3	8.3	4.3	4.0	4.1	4.0	4.0
Tergitol® 15-S-7	12.4	4.9	4.9	5.0	4.9	4.9
Tergitol® 15-S-9	13.3	3.5	3.6	3.5	3.5	3.5
Tergitol® 15-S-15	15.4	2.0	2.1	2.0	2.0	2.0
Tergitol® 15-S-20	16.3	1.5	1.6	1.5	1.4	1.5
Tergitol® 15-S-30	17.4	1.4	1.5	1.4	1.5	1.5
Tergitol® 15-S-40	18.0	1.1	1.2	1.2	1.2	1.3
Distilled water + triethanolamine carrier						
No surfactant		1.2	1.1	1.1	1.1	1.1
Tergitol® 15-S-3	8.3	3.6	3.8	4.5	4.0	4.0
Tergitol® 15-S-7	12.4	5.0	5.0	5.0	4.9	5.0
Tergitol® 15-S-9	13.3	3.6	3.4	3.5	3.5	3.5
Tergitol® 15-S-15	15.4	2.0	2.0	2.1	2.0	2.0
Tergitol® 15-S-20	16.3	1.5	1.5	1.5	1.5	1.5
Tergitol® 15-S-30	17.4	1.5	1.5	1.5	1.5	1.5
Tergitol® 15-S-40	18.0	1.2	1.2	1.2	1.2	1.2
LSD (5%)				0.2		

[1] HLB is hydrophilic/lipophilic balance value.
[2] AN (ammonium nitrate) at 0.25% (w/v), U (urea) at 0.25% (w/v), AN/U (ammonium nitrate/urea laboratory-made) at 0.25/0.25% (w/v), 28% N (commercial liquid nitrogen fertilizer) at 1% (v/v).
[3] 1 – low spread like distilled water, 5 – a total spread.

The pH of the spray mixture was increased more by TEA than 28% N (Tables 3 and 4). The higher pH from TEA than 28% N, which would enhance sulfosulfuron solubility, probably accounted for the greater sulfosulfuron efficacy. The use of TEA to attain a high pH greatly increased sulfosulfuron efficacy and reduced the importance of precise surfactant selection or even the need for nitrogen salts. The increased sulfosulfuron efficacy when TEA was included probably was from increased sulfosulfuron solubility at high pH rather than a direct effect of TEA on sulfosulfuron absorption, because TEA without surfactant only slightly increased sulfosulfuron efficacy (Table 6).

The TEA in the previous experiments was at 0.1% v/v of the spray mixture, which apparently was adequate to maximize sulfosulfuron efficacy with most surfactants (Table 6). The percentage TEA required to maximize efficacy with a surfactant decreased as a surfactant's efficacy increased. The pH of the spray mixture also increased with an increased TEA concentration (Table 7). For example, TEA at only 0.04% maximized sulfosulfuron efficacy when with the most effective Pluronic P 104® surfactant, but 0.16% concentration was required for the least effective Tergitol® 15-S-3. Sulfosulfuron

Table 6 – *Green foxtail fresh weight reduction from sulfosulfuron at 15 g ai/ha as influenced by secondary alcohol ethoxylate (Tergitol® 15-S) and block copolymer Pluronic® surfactants at 0.25% (v/v) and triethanolamine concentrations.*

Treatment	HLB[1]	Triethanolamine concentrations (v/v)					
		0	0.02	0.04	0.08	0.16	0.32
		% fresh weight reduction					
No surfactant		6	13	19	20	24	24
Tergitol® 15-S-3	8.3	22	24	36	39	45	43
Tergitol® 15-S-15	15.4	29	65	70	71	74	72
Tergitol® 15-S-40	18.0	26	64	81	83	84	88
LSD (5%)		4					
No surfactant		14	20	32	31	23	20
Pluronic® L 62	7.0	37	55	79	81	87	86
Pluronic® P 104	13.0	64	81	85	86	88	86
Pluronic® L 84	15.0	50	65	82	84	89	83
LSD (5%)		6					

[1] HLB is hydrophilic/lipophilic balance value.

Table 7 – *pH of sulfosulfuron (15 g ai/ha)* spray mixture *as influenced by secondary alcohol ethoxylate (Tergitol® 15-S) and block copolymer (Pluronic®) surfactants (0.25% v/v), and triethanolamine concentrations.*

Treatment	HLB[1]	Triethanolamine concentrations (% v/v)					
		0	0.02	0.04	0.08	0.16	0.32
		pH of spray mixture					
No surfactant		4.2	6.2	6.4	6.9	7.1	7.4
Tergitol® 15-S-3	8.3	4.1	6.1	6.3	7.0	7.1	7.4
Tergitol® 15-S-15	15.4	3.4	6.2	6.6	6.9	7.1	7.3
Tergitol® 15-S-40	18.0	5.2	6.3	6.6	6.9	7.1	7.3
No surfactant		4.1	6.2	6.6	6.9	7.2	7.4
Pluronic® L 62	7.0	4.1	6.1	6.5	6.8	7.1	7.2
Pluronic® L 104	13.0	4.0	6.1	6.5	6.8	7.2	7.4
Pluronic® L 84	15.0	4.0	6.2	6.4	7.0	7.2	7.3

[1] HLB is hydrophilic/lipophilic balance value.

applied with 0.16% TEA and Tergitol® 15-S-3 still only gave 45% fresh weight reduction compared to 85% with Pluronic® P 104 with only 0.04% TEA. Thus, pH or TEA alone does not maximize adjuvant efficacy of sulfosulfuron, but sulfosulfuron also requires the use of surfactants. Pluronic® surfactants generally were more effective than Tergitol® surfactants for sulfosulfuron. The greater Pluronic® effectiveness may reflect their larger molecular size as reported for long chain linear alcohol ethoxylate surfactants with nicosulfuron (Green 1999a).

High pH from TEA did not change droplet spread or drying time, which further indicated the direct importance of high pH to sulfosulfuron efficacy. Sulfosulfuron solubility in water increases from 18 to 1627 mg/L as pH is increased from 5 to 7 (Herbicide Handbook 1998). The high pH in the presence of TEA apparently dissolves the sulfosulfuron resulting in enhanced absorption.

These results with sulfosulfuron plus surfactants and nitrogen fertilizers differ from those with nicosulfuron (Nalewaja et al. 1995, Nalewaja et al. 1988, Nalewaja et al. 2000). Urea reduced nicosulfuron efficacy to large crabgrass [*Digitaria sanquinalis* (L.) Scop.] when applied with most oil adjuvants and nonionic surfactants. However, ammonium nitrate or 28% nitrogen fertilizer, but not urea, enhanced nicosulfuron efficacy to yellow foxtail when the spray water contained antagonistic sodium bicarbonate. In our experiments, addition of nitrogen fertilizers had a minor effect or was not important for sulfosulfuron efficacy, when efficacious surfactants were used and the spray mixture was maintained above pH 7. Thus, the benefit to herbicides from nitrogen fertilizers may depend upon the specific herbicide and associated surfactant or oil adjuvant. The common practice of including nitrogen salts with herbicides may not be beneficial with all herbicides and the benefit apparently depends on the specific nitrogen salt and accompanying characteristics imparted.

Conclusion

The results from these experiments demonstrate the importance of specific surfactant chemistry, surfactant HLB, and high pH for sulfosulfuron efficacy. Nitrogen fertilizers may not be necessary with spray carriers that have a high pH and an effective surfactant. However, with surfactants of unknown effectiveness, the use of nitrogen fertilizers in commercial practice would help assure sulfosulfuron efficacy especially when a high spray mixture pH is maintained.

Acknowledgment

This material is based upon work supported by the Cooperative States Research, Education, and Extention Service, U.S. Department of Agriculture, under Agreements No. 95-34330-1416 and 96-34361-2562.

References

Green J. M., 1999a, "Optimizing Alcohol Ethoxylate Surfactant Activity at Low Doses," *Weed Technology,* Vol. 13: 737-740.

Green, J. M., 1999b, "Effect of Nonylphenol Ethoxylation on the Biological Activity of Three Herbicides with Different Water Solubilities," *Weed Technology*, Vol. 13, pp. 840-842.

Gronwald, J. W., Jourdan, S. W., Wyse, D. L., Somers, D. A., and Magnusson, M. V., 1993, "Effect of Ammonium Sulfate on Absorption of Imezathapyr by Quackgrass," *Weed Science*, Vol. 41, pp. 325-334.

Herbicide Handbook, 1994, Ahrens, W. H., Ed., 7th Edition, Weed Science Society of America, Champaign, IL, pp. 216-217.

Herbicide Handbook, 1998, Hatzios, K. K., Ed., Supplement to 7th Edition, Weed Science Society of America, Champaign, IL, pp. 216-217.

Joost, R. E., 1998, "Benefits and Mode of Action of Nitrogen Fertilizers as Adjuvants," *Proc. Fifth International Symposium on Adjuvants for Agrochemicals*, Volume 1, pp. 259-266.

Lueschen, W. E., Hoverstad, T. R., and Gunsolus, J. L., 1991, "Wild Proso Millet Control in Corn," *Res. Rep. North Cent. Weed Sci. Soc.,* 48, pp. 236-237.

Manthey, F. A., Woznica, Z., and Milkowski, P., 1988, "Surfactants Differ in their Effect on Droplet Retention, Droplet Spread, and Herbicide Efficacy," *Pesticide Formulations and Application Systems: Eighteenth Volume, ASTM STP 1347*, J. D. Nalewaja, G. R. Goss, and R. S. Tann, Eds., American Society for Testing and Materials, pp. 120-130.

McWhorter, C. G., 1982, "The Use of Adjuvants," In *Adjuvants for Herbicides,* Hodgson, R. H., Ed. Weed Science Society of America, Champaign, IL, pp. 10-25.

Miller P. A., Westra P., and Nissen, S. J., 1999, "The Influence of Surfactant and Nitrogen on Foliar Absorption of MON 37500," *Weed Science*, Vol. 47, pp. 270-274.

Morton, C. A., Harvey, R. G., Anthon, J. M., and Kutil, J. L., 1991, "Wild-Proso Millet Control in Field Corn, Nicosulfuron Additives, Multistate Project," *Res. Rep. North Cent. Weed Sci. Soc.,* pp. 203-204.

Nalewaja, J. D. and Matysiak, R., 1993, "Spray Carrier Salts Affect Herbicide Toxicity to Kochia," *Weed Technology,* Vol. 7, pp. 154-158.

Nalewaja J. D., Matysiak, R., and Woznica, Z., 2000, "Optimum Surfactant HLB Value for Nicosulfuron is Salt Dependent," *Pesticide Formulations and Application Systems: Twentieth Volume, ASTM STP 1400,* A. K. Viets, R. S. Tann, and J. C. Mueninghoff, Eds., American Society for Testing and Materials, West Conshohocken, PA (in print).

Nalewaja, J. D., Praczyk, T., and Matysiak, R., 1995, "Salts and Surfactants Influence Nicosulfuron Efficacy," *Weed Technology*, Vol. 9, pp. 587-593.

Nalewaja, J. D., Praczyk, T., and Matysiak, R., 1998, "Nitrogen Fertilizer, Oil, and Surfactant Adjuvants with Nicosulfuron," *Weed Technology*, Vol. 12, pp. 585-589.

Woznica, Z., Nalewaja, J. D., and Szelezniak, E. F., 1997, "MON 37532 Phytotoxicity is Affected by Surfactant and Ammonium Nitrate," *Pesticide Formulations and Application Systems: Seventeenth Volume, ASTM STP 1328,* R. G. Goss, M. J. Hopkinson, and H. M. Collins, Eds., American Society for Testing and Materials, Philadelphia, PA, pp. 287-297.

Hans de Ruiter,[1] Roger A. Downer,[2] André J. M. Uffing,[1] Timothy A Ebert,[2] Pieter J. J. Pikaar,[1] and Frank R. Hall[2]

The Influence of Inorganic Cations on Glyphosate Activity-Review and Perspectives

---

Reference: De Ruiter, H., Downer, R. A., Uffing, A. J. M., Ebert, T. A., Pikaar, P. J. J., and Hall, F. R., "**The Influence of Inorganic Cations on Glyphosate Activity- Review and Perspectives,**" *Pesticide Formulations and Application Systems: A New Century for Agricultural Formulations, Twenty First Volume, ASTM STP 1414,* J. C. Mueninghoff, A. K.Viets, and R. A. Downer, Eds., American Society for Testing and Materials, West Conshohocken, PA, 2001.

Abstract: The influence of calcium, iron, magnesium, other cations, and mixtures of cations on the activity of glyphosate and the interaction with pH is reviewed, discussed, and demonstrated with laboratory data. The strategy to overcome antagonistic effects of the cations via use of adjuvants is discussed and illustrated with data.

The influence of calcium concentration on glyphosate efficacy was investigated on black nightshade (*Solanum nigrum* L.) by monitoring glyphosate activity via fluorescence technology and inhibition of growth. Giant foxtail (*Setaria faberi* Herrm) was also used for biological evaluation. Calcium concentrations equal to the glyphosate concentrations were most antagonistic; higher concentrations of calcium were less antagonistic.

Different adjuvants were tested to overcome the antagonistic effect of calcium. Results are discussed in relation to the physical chemical properties of the adjuvants. With a more appropriate scientific base, remedial measures may protect field performance of glyphosate and reduce farm costs.

Keywords: glyphosate, cations, calcium, antagonism, adjuvant, ammonium sulfate, surfactant

Presence of inorganic ions like Ca^{2+}, Mg^{2+}, Zn^{2+}, Fe^{2+}, Fe^{3+}, and Al^{3+} ions in the carrier liquid can reduce the activity or foliar uptake of glyphosate (Buhler and Burnside

---

[1] Senior Scientist and Research Assistants, Plant Research International, P. O. Box 16, NL-6700 AA Wageningen, The Netherlands.

[2] Research Scientist, Research Associate, and Professor Emeritus, Laboratory for Pest Control Application Technology (LPCAT), The Ohio State University, Wooster, OH, 44691, USA.

1983, Gauvrit et al. 2001, Hensley et al. 1978, Nalewaja and Matysiak 1991, Nalewaja et al. 1996, Thelen et al. 1995a, Wills and McWhorter 1985). Furthermore, calcium ions, either on the leaf surface or within the leaf tissue, may also reduce the foliar uptake of glyphosate (Hall et al. 2000). At a pH>4 , the dissociation of the carboxylic and the phosphate group give the glyphosate molecule negative charges. The ion species distribution (mono-, di- and trianion) is pH-dependent (Wauchope 1976). The interaction of glyphosate with inorganic cations can result in water-soluble salts (for example K^+ and Na^+ salts) with a high herbicidal activity and in water-insoluble salts or complexes (Ca^{2+}, Mg^{2+}, Fe^{3+}, and Al^{3+}-salts), which are less effective (Franz et al. 1997).

With regard to calcium, an NMR study indicated the presence of 1:1 (glyphosate:calcium) complexes in dilute solutions (Thelen et al. 1995a). A time-course experiment on the foliar uptake of glyphosate (Roundup® formulation) indicated that most foliar uptake occurs after visible drying of the drop (Gaskin and Stevens 1993). Calcium in the carrier created a more amorphous-appearing deposit with poorer contact with the leaf surface, which was thought to hinder foliar uptake (Nalewaja et al. 1992). At this stage, it is not known, whether the calcium-induced immobilization of glyphosate in the dried deposit differs from the glyphosate calcium association in dilute solutions. An indirect approach whereby different calcium concentrations were tested at a range of glyphosate doses provided evidence that the glyphosate calcium antagonism is based on a 1:1 glyphosate:calcium complexation (Gauvrit et al. 2001).

In this paper we focused our efforts on a number of questions that are considered as relevant to handle the antagonism under field conditions. In water used as a carrier for crop protection agents, calcium is one of the most frequently occurring inorganic cations (Buhler and Burnside 1983, Nalewaja and Matsyak 1993, Shea and Tupy 1984, Stahlmann and Phillips 1979). Concentrations of 500 ppm (12.5 mM) (Nalewaja and Matsyak 1993) and 850 ppm (Haire et al. 1985a) have been reported. However, in most situations, the calcium concentration varies between 50 and 200 ppm (1.25-5 mM) (Anonymous 1998, Buhler and Burnside 1983, Haire et al. 1985b, Nalewaja and Matsyak 1993, Shea and Tupy 1984, Stahlman and Phillips 1979). Consequently, most of our attention is focused on the influence of calcium-ions. After discussing the influence of pH, we investigated the present state of the art with regard to the influence of specific inorganic cations and the influence of mixtures of these ions. In relation to calcium, we discuss how the ratio between glyphosate and calcium in the carrier liquid influences the antagonism. In this respect, we addressed the question how an adjuvant changes the relation between glyphosate/calcium ratio and the antagonism. We also discuss the several options already available, or to be developed, in overcoming the calcium glyphosate antagonism. The review is illustrated with research data from the laboratories of the authors using black nightshade (*Solanum nigrum* L.) and giant foxtail (*Setaria faberi* Herrm) as test species.

Influence of pH on Calcium-Glyphosate Antagonism

The pH-dependent ion distribution of glyphosate (Wauchope 1976) looks like a reasonable argument to suggest that the degree of complexation of calcium with glyphosate depends on the pH of the treatment solution. An earlier report (Madsen et al. 1978) indicated that calcium will complex with glyphosate at alkaline pH (pH>8). This should mean that particularly the glyphosate tri-anion is involved in the glyphosate-calcium association. NMR data indicate an association of calcium with both the carboxyl and phosphonate groups on the glyphosate molecule (Thelen et al. 1995a). Both groups are negatively charged at a broad pH range, which indicates that the glyphosate-calcium complexation also occurs at a broad range of pH values. A thorough and carefully designed study (Nalewaja and Matysiak 1991) suggested that not pH but the concentration and the type of anionic counter-ion used to buffer the glyphosate solution determined the level of glyphosate-calcium antagonism. The degree of antagonism was constant from pH 2 to pH 9 with NO_3^- as the anion but was much less pronounced at pH<6 when PO_4^{3-} or SO_4^{2-} were included as the anion. We also observed (Plant Research International, unpublished data) that the glyphosate-calcium antagonism was observed at pH values around 5 which is not in line with the indication from Madsen et al. (1978) that the association will be observed at pH>8. Thus it seems reasonable to conclude that the glyphosate-calcium antagonism is not influenced significantly by pH. It can be reasoned that the pH of the carrier liquid will have little influence on the efficacy of glyphosate when the carrier liquid does not contain mineral salts. The log Ko/w of glyphosate is –4.3, -4.8, and –4.8 at pH 3.0, 5.6, and 8.0 respectively (partly published in de Ruiter and Meinen 1996). This indicates that passive diffusion of glyphosate through lipohilic barriers like the leaf cuticle and the cell membrane is not pH dependent. We (LPCAT, unpublished data) observed that reducing the pH from 4.9 to 1.1 with hydrochloric acid did not affect glyphosate efficacy.

Influence of Different Inorganic Cations on Glyphosate Phytotoxicity

Previous reports provide data (Table 1) on glyphosate-cation antagonism that were collected under different conditions with regard to concentration of the cation and its counter-ion, the concentration of glyphosate, the pH of the test solution and the test species plus test method.

Table 1-*Reported influence of different inorganic cations on glyphosate activity*[1]

	Inorganic cations								
Reference	Ca^{2+}	Mg^{2+}	Zn^{2+}	Fe^{2+}	Fe^{3+}	Al^{3+}	K^+	Na^+	NH_4^+
Buhler and Burnside 1983	S	M	S	M	S			N	
De Villiers et al. 1996	S	M							
Hensley et al. 1978	N			S		S	N	N	
Nalewaja and Matysiak 1991	S	S	S	S	S			M	N
Nalewaja and Matysiak 1993	S	M							
Stahlman and Phillips 1979	S	M	S	S		S	N	N	
Wills and McWhorter 1985	N		S		S		E	E	E

[1]S=Severe reduction of glyphosate activity (>50%); M=Moderate reduction (<50%); N=No influence; E=Enhancement of glyphosate activity.

In spite of these differences we draw the following generalization from the data (Table 1). The generalization is not complete due to the fact that not all inorganic cations have been tested with glyphosate or their data were not included in this paper. The group of ions reducing glyphosate activity by more than 50% includes: Ca^{2+},Zn^{2+}, Fe^{2+}, Fe^{3+}, and Al^{3+}. The Mg^{2+}-ions are moderate inhibitors of glyphosate activity and Na^{+} and K^{+} do not strongly influence glyphosate activity. It is interesting to see that two reports place calcium ions in the group of inert cations. Wills and McWhorter (1985) applied calcium ions at a relatively high concentration of 100 mM (4000 ppm) and Hensley et al. (1978) tested calcium ions on glyphosate activity via a root bioassay using a mix of sand and glyphosate solution as a medium. In a later section (Influence of Glyphosate/Calcium Ratio on the Antagonism) we discuss these remarkable data on calcium-ions.

The generalization as described above agrees with measurements of the solubility products of glyphosate-metal salts and their biological activity using Ca^{2+}, Mg^{2+},Zn^{2+}, Fe^{3+}, Mn^{2+}, and Cu^{2+} as cations (Sundaram and Sundaram 1997). These authors concluded that the larger the solubility product, the greater the concentration of glyphosate ions in solution and the greater the growth inhibition.

Influence of Mixtures of Different Inorganic Cations on Glyphosate Phytotoxicity

Water used as a carrier for application of glyphosate commonly contains a mixture of inorganic cations. Several of the previously cited studies investigated the influence of individual cations using distilled water, but these experiments do not indicate ion interaction between the cations when present as a mix in the carrier. A step in finding interactions was done by Nalewaja and Matysiak (1993) by testing different ratios of $CaCl_2$ and $NaHCO_3$. Although not subjected to a regression model, their data indicated that the antagonistic effects of these salts were additive. Both the cation and the anion varied thus it was not possible to make a statement on interactions between individual ions. Therefore, a limited number of mixtures were tested (LPCAT) for interactions between Ca^{2+}, Fe^{3+}, and Cu^{2+}.

Table 2-*Mixtures of copper-, iron- and calcium-ions tested with glyphosate on foxtail.*[1]

		Mixture number									
		1	2	3	4	5	6	7	8	9	10
Experiment	Cation	Concentration of cation (mM)									
1	Cu^{2+}	6.7	0	20	0	13.3	13.3	6.7	6.7	0	0
	Fe^{3+}	6.7	0	0	20	6.7	0	13.3	0	6.7	13.3
	Ca^{2+}	6.7	20	0	0	0	6.7	0	13.3	13.3	6.7
2	Cu^{2+}	1.7	0	5	0	3.3	3.3	1.7	1.7	0	0
	Fe^{3+}	1.7	0	0	5	1.7	0	3.3	0	1.7	3.3
	Ca^{2+}	1.7	5	0	0	0	1.7	0	3.3	3.3	1.7

[1]Included in a distilled water carrier with Roundup-Ultra (13.7 mM); Experiment 1 was conducted one time, experiment 2 was conducted twice.

Three chloride salts of these metals were tested at constant ionic strength and chloride concentration (Table 2). Experiments with ten different mixtures of the three cations were conducted whereby the total molar concentration (sum of three individual cations) is 20 mM (Experiment 1) or 5 mM (Experiment 2). The salts were included in a Roundup-Ultra solution (13.7 mM) and applied on giant foxtail.

Regression analysis of the plant fresh weights showed no interactions between iron, copper, and calcium. This means that glyphosate antagonism effects of iron, copper, and calcium are simply additive in spite of the observation that the solubility product of the 1:1 glyphosate-calcium salt is several orders of magnitude higher than that of the copper and iron salt of glyphosate (Sundaram and Sundaram 1997). In the bioassay on giant foxtail, glyphosate moves from the visually dry drop deposits into the plant. The study cited with regard to the solubility product (Sundaram and Sundaram 1997), used a glyphosate solution in which the roots of tomato plants were immersed. We suggest that the mobilization of glyphosate on a leaf surface is more complex and this may have obscured the differences as expected on the solubility products of the metal glyphosate salts. We suggest that the situation on the leaf surface makes the iron and copper ions less antagonistic.

The Influence of Glyphosate/Calcium ratio on the Calcium Antagonism

Most previous studies on the calcium antagonism of glyphosate expressed the glyphosate application rates in g ae/ha and the calcium concentration in the carrier in ppm or molarity. From the physical chemical point of view it is obvious that the molar concentration of both the glyphosate ions and the calcium ions determine the fraction of glyphosate inactivated by complexation with calcium. The NMR study of Thelen et al. (1995a) and the indirect approach by Gauvrit et al. (2001) indicate that the calcium antagonism of glyphosate is based on a 1:1 glyphosate/calcium complexation.

Table 3-*The influence of the glyphosate/calcium ratio on the calcium antagonism of glyphosate.*

	Degree of calcium antagonism[1,2,3]		
Glyphosate/calcium ratios	S[4]	M[4]	N[4]
Mean value (mM/mM)	1.1	6.0	8.3
Standard deviation (SD)	1.0	3.9	4
Number of values (N)	23	4	8

[1]Using fresh weight or dry weight reduction as parameter for glyphosate activity.
[2]$CaCl_2$ or $Ca(OH)_2$ (Nalewaja et al. 1996) was used as source for Ca^{2+}-ions.
[3]Studies included: Buhler and Burnside 1983, de Villiers et al. 1996, Nalewaja and Matysiak 1991, Nalewaja and Matysiak 1992, Nalewaja et al. 1996, Shea and Tupy 1984, Shilling and Haller 1989, Stahlmann and Phillips 1979.
[4]S=Severe reduction (>50%); M=Moderate reduction (<60%); N=no reduction

We investigated (Table 3) whether this ratio is also valid for the previously reported calcium antagonism of glyphosate efficacy and glyphosate uptake. If a similar relationship is found, then the glyphosate/calcium ratio on a molar basis may be one of the parameters that predict the occurrence of antagonism of glyphosate.

The glyphosate/calcium ratios around 1 result in a severe calcium-antagonism and values of 6 and higher give a moderate or no reduction of glyphosate efficacy (Table 3). The foliar uptake of glyphosate was also reduced severely by calcium ions when the glyphosate/calcium ratio was around 1; values of 1.4 (Gauvrit et al. 2001), 0.3 (Thelen et al. 1995a), and 1 (Nalewaja et al. 1996) could be calculated.

Throughout the literature search, both severe calcium-antagonism of glyphosate efficacy and absence of antagonism has been observed at a ratio of 3. In the first situation (severe antagonism at ratio of 3) the glyphosate application rate is such that glyphosate alone is just sufficient and as a consequence partial inactivation by calcium will easily impact the efficacy. In the second situation (no antagonism at ratio of 3) the glyphosate dose applied is more than required for good efficacy and a partial inactivation due to calcium will not reduce the activity.

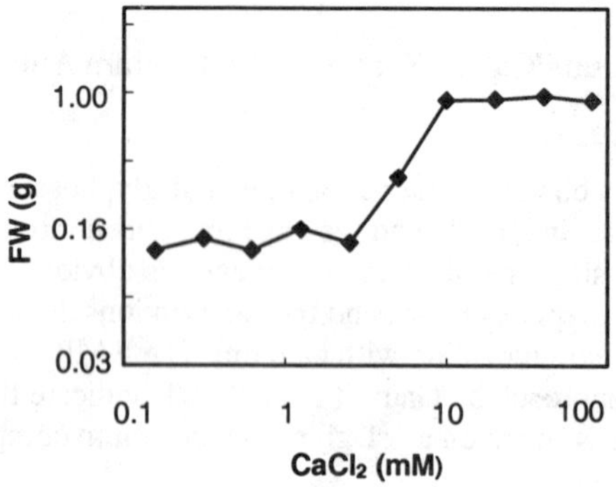

Figure 1-*Influence of calcium in the carrier liquid on the efficacy of glyphosate (Roundup Ultra®) applied on giant foxtail at 14 mM. Fresh weight of untreated plants was 1.0 g. Data (LPCAT, Ohio) are mean values of two separate experiments each with six replicates.*

The relevance of the glyphosate/calcium ratio on a molar basis results in the following insights. It is understandable that lowering of the carrier volume can reduce the calcium antagonism (Buhler and Burnside 1983, Sandberg et al. 1978, Stahlman and Phillips 1979) because reduction of carrier volume proportionally increases the ratio. Application of glyphosate at the recommended rates results in a glyphosate concentration varying approximately between 20 and 60 mM. This implies that carrier water with 1-5 mM calcium (40-200 ppm) will not seriously impact glyphosate efficacy unless the recommended rate is just sufficient in a certain situation (weather conditions, weed susceptibility). However in all those situations where glyphosate is applied at concentrations around 5mM (carrier volumes>1000 L/ha, low dose systems on young weed seedlings, mixtures with other active ingredients) or where the calcium concentration is extremely high [10 mM (400 ppm)or higher], there we can expect reduction of efficacy due to calcium plus the other antagonistic ions (Mg^{2+}, Zn^{2+}, Fe^{3+}, Mn^{2+}, and Cu^{2+}) if present in the carrier.

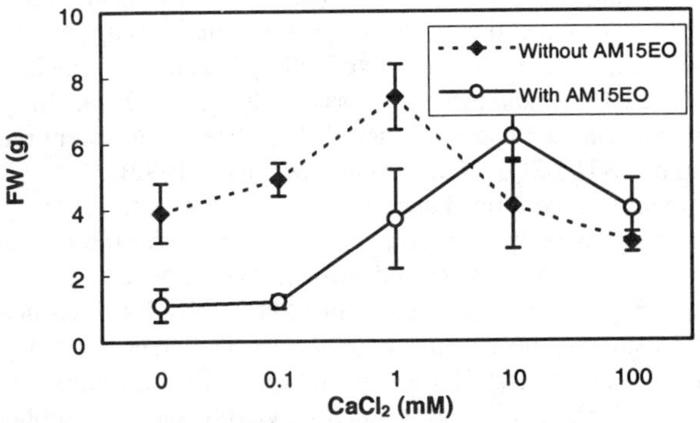

Figure 2- *Influence of calcium in the carrier liquid on the efficacy of unformulated glyphosate applied on giant foxtail at 1 mM. Solutions were applied without and with a polyoxyethylene (15) tallowamine (AM15EO; Ethomeen® T/25[3]) at 0.1% (w/v). Fresh weight of untreated plants was 10 g. Data (Plant Research International, The Netherlands) are mean values of a representative experiment out of two separate experiments each with four replicates.*

Experiments on giant foxtail (Figure 1) and black nightshade (Figure 2) confirm that equal molar concentrations of calcium and glyphosate may result in a severe antagonism. The data with black nightshade (Figure 2) indicate that the AM15EO surfactant gives some but not a complete protection against calcium. Inclusion of calcium at 100 mM reduced the antagonism strongly without AM15EO and less strongly with AM15EO. A reduction of the calcium antagonism was also observed previously at 40-800 mM Ca^{2+}(Stahlman and Philips 1979) and 100 mM Ca^{2+}(Wills and McWhorter 1985). We suggest that an excess of the hygroscopic $CaCl_2$ may enhance the wetting of the drop deposit such that glyphosate ions can move into the plants.

Use of Adjuvants to Overcome the Calcium Antagonism of Glyphosate

It is a challenge to develop a glyphosate formulation of which the efficacy is not reduced by antagonism caused by calcium or other metals. This enables the applicator to

[3] Akzo Nobel, Stenungsund, Sweden.

reduce glyphosate application rates. The following compounds are reported to be effective against the calcium antagonism: $(NH_4)_2SO_4$ (Nalewaja and Matysiak 1993, Shiller et al. 1990, Thelen et al. 1995), $(NH_4)_2SO_4$ + surfactant (Nalewaja and Matysiak 1992), phosphate ions (Nalewaja and Matysiak 1991, Shilling et al. 1990), sulfate ions (Buhler and Burnside 1983, Nalewaja and Matysiak 1991), citrate (Nalewaja and Matysiak 1991, Thelen et al. 1995b), and EDTA (Shea and Tupy 1984). On a molar basis, a surplus of these adjuvants is required compared to the concentration of calcium in the carrier. $(NH_4)_2SO_4$ is an exception in this respect. Calcium-antagonism of glyphosate activity was neutralized at equimolar amounts of $(NH_4)_2SO_4$ in an application on wheat (Nalewaja and Matysiak 1992) and torpedo grass (Shiller et al. 1990). On species like kochia and soybean, a surfactant had to be included to observe sufficient reduction of the calcium antagonism by $(NH_4)_2SO_4$ (Nalewaja and Matysiak 1992)

Most of the compounds mentioned above (sulfate, phosphate, citrate, and EDTA) reduce the calcium-antagonism by competing with glyphosate for the formation of more or less insoluble glyphosate salts (Buhler and Burnside 1983, Nalewaja and Matysiak 1991, Thelen et al 1995a, Thelen et al. 1995b). In addition NH_4^+ ions compete with Ca^{2+} ions for complexation sites on the glyphosate molecule (Thelen et al 1995a).

In a two-factorial design $(NH_4)_2SO_4$ was tested as an adjuvant to overcome the calcium antagonism of glyphosate (Figures 3 and 4). Glyphosate was applied in a suboptimal concentration (1 mM) as a spray (Figure 3) and at a more common concentration of 20 mM in two 1-μl drops on the first leaf (Figure 4). In both situations, Ca^{2+} was most antagonistic at equal molar concentrations of glyphosate and Ca^{2+}, which agrees with the previous discussion. $(NH_4)_2SO_4$ overcame the calcium antagonism when included at the same concentration of calcium (Figures 3 and 5) or gave a further reduction of fresh weight. (Figure 3).

Ammonium sulfate can also enhance glyphosate activity when applied in distilled or demineralized water on sunflower (Nalewaja et al. 1992) and wheat (de Ruiter et al. 1994, Nalewaja and Matysiak 1991). Experiments with a range of species (unpublished data) demonstrated this enhancement at a concentration of 60 to 80 mM $(NH_4)_2SO_4$ (7.5-10 g/L). It cannot be excluded, that at least part of the glyphosate activator function in calcium-free water is caused by "inactivation" of calcium present on the leaf surface and in the leaf cuticle as demonstrated and discussed previously (Hall et al. 2000). Calcium chelating agents may also have a function in the plant. Most apoplastic calcium (98%) is bound to the cell walls leaving a relatively low concentration of 0.3-0.8 mM in the apoplastic fluid (Mühling and Sattelmacher 1995). Removal of calcium from the cell walls and cell membranes by calcium chelating agents like sulfate ions may increase the cellular uptake of glyphosate. An $(NH_4)_2SO_4$ enhanced uptake of glyphosate into quackgrass protoplasts has been observed (de Ruiter and Meinen 1996). Further, we suggest that $(NH_4)_2SO_4$ causes a locally high ionic strength which reduces the ionic binding of calcium with pectate in the cell-wall.

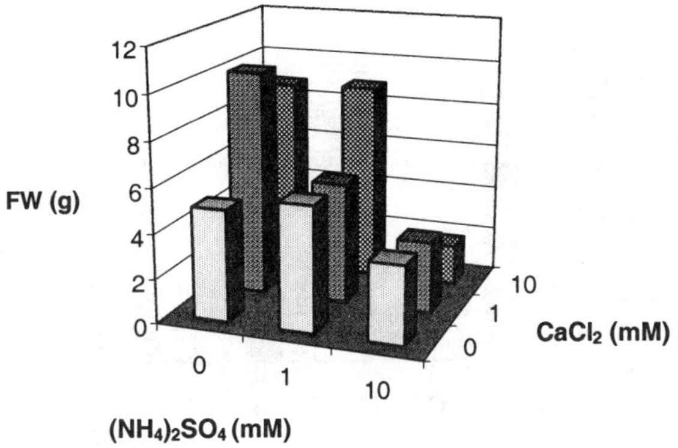

Figure 3-*Influence of calcium and ammonium sulfate in the carrier liquid on the efficacy of unformulated glyphosate when applied at 1 mM as a spray on black nightshade. The fresh weight of untreated plants was 11 g. Data (Plant Research International, The Netherlands) are mean values of two separate experiments each with four replicates.*

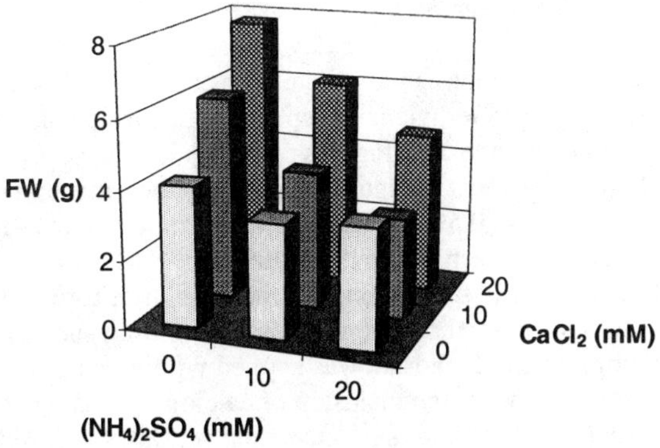

Figure 4-*Influence of calcium and ammonium sulfate in the carrier liquid on the efficacy of unformulated glyphosate when applied at 20 mM as two 1-µl drops on the first leaf of black nightshade plants 94-5 leaves). The fresh weight of untreated plants was 12.9 g. Data (Plant Research International, The Netherlands) are mean values of two separate experiments each with four replicates.*

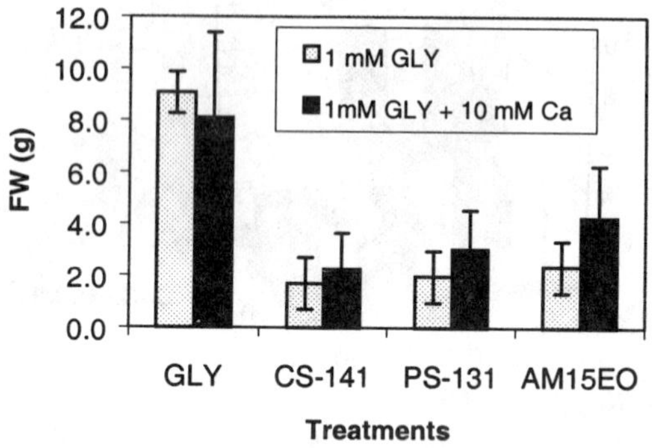

Figure 5-*Influence of a polyoxyethylene (10) nonylphenol phosphate ester (Emphos® CS-141*[4]*), a polyoxyethylene (6) tridecyl alcohol phosphate ester (Emphos® PS-131*[4]*) and a polyoxyethylene (15) tallowamine (AM15EO) [all at 0.5% (w/v)] on the calcium antagonism (10 mM $CaCl_2$) of unformulated glyphosate (GLY) at 1 mM. Solutions were applied as a spray on black nightshade. Fresh weight of the untreated plants was 9.25 g. Data (Plant Research International, The Netherlands) are mean values of two separate experiments each with four replicates.*

Charged phosphate groups can associate with calcium ions. For that reason we investigated whether polyoxyethylene phosphate ester surfactants can overcome the calcium antagonism of glyphosate (Figures 5, 6, and 7). The phosphate ester surfactants enhanced the glyphosate activity (Figures 5 and 6) to the same extent as the AM15EO surfactant when included at 0.5% (w/v) in the carrier liquid. $CaCl_2$ in the solutions at 10 mM (400 ppm) reduced the activity in the presence of AM15EO but this reduction was absent or less less obvious in the presence of the phosphate ester surfactants (Figures 5 and 6). A very low sub-optimal rate of glyphosate was used such that there was no influence of calcium when glyphosate was applied without surfactant. Foliar uptake experiments demonstrated that the inclusion of calcium ions did not reduce the uptake in the presence of the phosphate ester surfactants but seriously reduced the uptake with AM15EO. In the uptake experiment the surfactants were included at 0.1% (w/v) and this may explain the lower uptake enhancement as was expected from the activity data (Figures 5 and 6). Further experiments indicated that, in comparison to the AM15EO surfactant, a relatively high concentration of the phosphate ester surfactants (0,5 %) is required to enhance substantially the glyphosate activity.. The concentration of phosphorus in the surfactants tested is such that 6 to 7 mM was present in the treatment solution with 0.5% surfactant and 1.2 to 1.4 mM at 0.1%[5]. This is lower than the 10 mM

[4] Witco-Crompton Corporation, Greenwich, USA.

[5] Personal communication, Dennis Anderson, Witco-Crompton Corp., October 2000.

Ca^{2+}, which indicates the potential of these products to overcome the calcium antagonism.

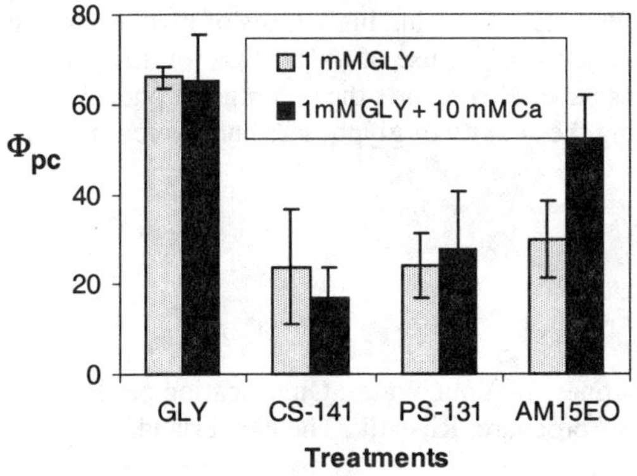

Figure 6-*Same experiments as Figure 5. Accumulation of glyphosate (GLY) in a young sink leaf was monitored using fluorescence technology (de Ruiter et al. 1998). The parameter ϕ_{pc} is inversey proportional to the amount of glyphosate entering the young sink leaf.*

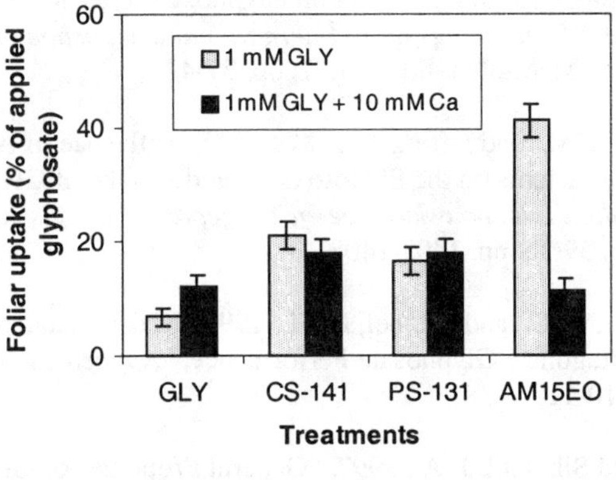

Figure 7-*Influence of a polyoxyethylene (10) nonylphenol phosphate ester, a polyoxyethylene (6) tridecyl alcohol phosphate ester and a polyoxyethylene (15) tallowamine (AM15EO) (all at 0.1%) on the foliar uptake of unformulated glyphosate (GLY) at 1 mM into black nightshade. Uptake was measured without and with 10 mM $CaCl_2$. Data (Plant Research International, The Netherlands) are mean values of three separate experiments each with four replicates.*

Conclusions

All situations where the molar glyphosate/calcium ratio or the molar ratio between glyphosate and the sum of copper, iron and calcium ions is around one, antagonism of glyphosate may result. An alternative for using high doses of glyphosate is to include adjuvants that reduce the antagonism. The use of $(NH_4)_2SO_4$ for this purpose is effective and is well documented. This paper also reports the potential of phosphate ester surfactants for both enhancing the activity of glyphosate and overcoming the calcium antagonism.

References

Anonymous, 1998, "The Hardness of Your Water; Qualification per Region in The Netherlands," VEWIN brochure, Rijswijk, The Netherlands.

Buhler, D. D. and Burnside, O. C., 1983, "Effect of Water Quality, Carrier Volume, and Acid on Glyphosate Phytoxicity," *Weed Science*, Vol. 31, pp. 163-169.

De Ruiter, H. and Meinen, E., 1996, "Adjuvant-Increased Glyphosate Uptake by Protoplasts Isolated from Quackgrass *Elytrigia repens* (L.) Nevski," *Weed Science*, Vol. 44, pp. 38-45.

De Ruiter. H., Meinen, E., Ribot, S. A., and Uffing, A. J. M., 1998, "A New Approach for Monitoring the Influence of Adjuvants on Glyphosate Transfer from Drop Deposit to Sink Leaf," *Proceedings of 5th International Symposium on Adjuvants for Agrochemicals*, P. McMullan, Ed., Vol. 1, pp. 37-42.

De Ruiter, H, Verbeek, M. A. M., and Uffing A. J. M., 1994, "Influence of Ammonium Sulphate and two Surfactants on the Phytotoxicity and Uptake of Glyphosate," *Mededelingen Faculteit Landbouwkundige en Toegepaste Biologische Wetenschappen*, No. 59/3b, pp. 1403-1408.

De Villiers, B. L., Lindeque, R. C., and Knobel, H. H., 1996, "Calcium and Magnesium in Spray Carriers antagonize Glyphosate Performance," *Applied Plant Science*, Vol. 10, No. 1, pp. 10-11

Franz, J.E., Mao, M. K., and Sikorski, J. A., 1997, "General Properties of Glyphosate and Glyphosate Salts," *Glyphosate, A Unique Global Herbicide*, ACS Monograph 189, American Chemical Society, Washington, DC, pp.27-64.

Gaskin, R. E., and Stevens, P. J. G., 1993, "Antagonism of the Foliar Uptake of Glyphosate into Grasses by Organosilicone Surfactants. Part 1: Effects of Plant

Species, Formulation, Concentrations and Timing of Application," *Pesticide Science*, Vol. 38, pp. 185-192.

Gauvrit, C., Gaudry, J. C., Lucotte, T., and Cabanne, F., 2001, "Evidence for a 1:1-Ca^{2+}: Glyphosate Association in Deposits Residuals on Leaf Surface," Proceedings 52st International Symposium on Crop Protection, *Mededelingen Faculteit Landbouwkundige en Toegepaste Biologische Wetenschappen*, In press.

Hall, G. J., Hart, C. A., and Jones, C. A., 2000, "Plants as Sources of Cations Antagonistic to Glyphosate Activity," *Pest Management Science*, Vol. 56, p. 31-358.

Haire, W. J., Price, C, Curtis, R. E. and Lietz, C., 1985a, "Water Resources Data, Florida, Water year 1985," Vol. 2A, *South Florida Surface Water*, United States Geological Survey, Miami, Florida.

Haire, W. J., Sonenshein, R., and Price, 1985b, "Water Resources Data, Florida, Water year 1985," Vol. 2B, *South Florida Groundwater*, United States Geological Survey, Miami, Florida.

Hensley, D. L., Beuerman, D. S. N., and Carpenter, P. L., 1978, "The Inactivation of Glyphosate by Various Soils and Metal Salts," Weed Research, Vol. 18, pp. 287-291.

Madsen, H. E. L., Christensen, H. H., and Gottlieb-Petersen, C., 1978, "Stability Constants of Copper (II), Zinc, Manganese (II), Calcium, and Magnesium Complexes of *N*-(Phosphonomethyl)glycine (Glyphosate)," *Acta Chemica Scandinavica*, Vol. A32, No. 1, pp. 79-83.

Mühling, K. H. and Sattelmacher, B., 1995, "Apoplastic Ion Concentration of Intact Leaves of Field Bean (Vicia faba) as influenced by Ammonium and Nitrate Nutrition," *Journal of Plant Physiology*, Vol. 147, pp. 81-86.

Nalewaja, J. D. and Matysiak, R., 1991, "Salt Antagonism of Glyphosate," *Weed Science*, Vol. 39, pp. 622-628.

Nalewaja, J. D. and Matysiak, R., 1992, "Species differ in Response to Adjuvants with glyphosate," *Weed Technology*, Vol. 6, pp. 561-566.

Nalewaja, J. D. and Matysiak, R., 1993, "Optimizing Adjuvants to Overcome Glyphosate Antagonistic Salts," *Weed Technology*, Vol. 7, pp. 337-342.

Nalewaja, J. D., De Villiers, B., and Matysiak, R., 1996, "Surfactant and Salt affect Glyphosate Retention and Absorption," *Weed Research*, Vol. 36, No. 3, pp. 241-248.

Nalewaja, J. D., Matysiak, R., and Freeman, Th. P., 1992, "Spray Droplet Residual of Glyphosate in Various Carriers," *Weed Science*, Vol. 40, pp. 576-589.

Sandberg, C. L., Meggitt, W. F., and Penner, D., 1978, "Effect of Diluent Volume and Calcium on Glyphosate Phytotoxicity," *Weed Science*, Vol. 26, pp. 476-479.

Shea, P. J., and Tupy, D. R., 1984, "Reversal of Cation-Induced Reduction in Glyphosate Activity with EDTA," *Weed Science*, Vol. 32, pp. 802-806.

Shilling, D. G. and Haller, W. T., 1989, "Interactive Effects of Diluent pH and Calcium Content on Glyphosate Activity on *Panicum repens* L. (torpedograss)," *Weed Research*, Vol 29, pp. 441-448.

Shilling, D. G., Haller, W.T., Willard, T. R., and Mossler, M. A., 1990, "Influence of Surfactants and Additives on Phytotoxicity of Glyphosate to Torpedogras," *Journal of Aquatic Plant Management*, Vol. 28, pp. 23-27.

Stahlman, P.W. and Philips, W.M., 1979, "Effects of Water Quality and Spray Volume on Glyphosate Phytotoxicity," *Weed Science*, Vol. 27, 38-41

Sundaram, A. and Sundaram, K. M. S., 1997, "Solubility products of Six Metal-Glyphosate Complexes in Water and Forestry Soils, and their Influence on Glyphosate Toxicity on Plants," *Journal of Environmental Science and Health*, Vol. B32, No. 4, pp. 583-598.

Thelen, K., D., Jackson, E. P., and Penner, D., 1995a, "The Basis for the Hard-Water Antagonism of Glyphosate Activity," *Weed Science*, Vol. 43, pp. 541-548.

Thelen, K. D., Jackson, E. P., and Penner, D., 1995b, "Utility of Nuclear Magnetic Resonance for Determining the Molecular Infleunce of Citric Acid and an Organosolicone Adjuvant on Glyphosate Activity," *Weed Science*, Vol. 43, pp. 566-571.

Wauchope, D., 1976, "Acid Dissociation Constants of Arsenic Acid, Methylarsonic Acid (MAA), Dimethylarsinic Acid (Cacodylic Acid), and *N*-(Phosphonomethyl)glycine (Glyphosate)," *Journal of Agricultural and Food Chemistry*, Vol. 24, No. 4, pp. 717-721.

Wills, G. D. and McWhorter, C. G, 1985, "Effect of Inorganic Salts on the Toxicity and Translocation of Glyphosate and MSMA in Purple Nutsedge (*Cyperus rotundus*)," *Weed Science*, Vol. 33, pp. 755-761.

SESSION II: Methods

R. Scott Tann,[1] P. Douglas Frisch,[2] and Karen P. Kuo[2]

Pesticide Solubility and Other Tools to Use to Optimize an Emulsifiable Concentrate Formulation

---

Reference: Tann, R. S., Frisch, P. D., Kuo, K. I., **"Pesticide Solubility and Other Tools to Use to Optimize an Emulsifiable Concentrate Formulation,"** *Pesticide Formulations and Application Systems: A New Century for Agricultural Formulations, Twenty First Volume, ASTM STP 1414*, J. C. Mueninghoff, A. K. Viets, and R. A. Downer, Eds., American Society for Testing and Materials, West Conshohocken, PA, 2001.

Abstract: Emulsifiable concentrates (EC) continue to be the preferred formulation type as judged by the number of new formulations registered each year. The key component of an EC formulation is a solvent capable of dissolving the active ingredient. This paper examines common types of hydrocarbon and oxygenated solvents registered for agricultural chemical use and identifies some critical physical properties that can be used to guide the formulator in his solvent selection. Properties such as Kauri Butanol (KB) Value (ASTM D1133), Aniline Point (ASTM D611), and Hildebrand or Hansen Solubility Parameter are generally recognized, direct measures of solvency. If applied properly, these properties can directionally give valuable starting information on solvency. Each of these test methods will be described together with how it relates to solvency of a fluid. In addition, the relevance to the formulator of ECs will be demonstrated by correlation of the results of these tests with laboratory-determined solubility of several generic pesticides.

Keywords: emulsifiable concentrate, solubility, aniline point, Kauri-Butanol value, Hildebrand Solubility

---

[1]Staff Market Planner, Intermediates, ExxonMobil Chemical Company, 13501 Katy Freeway, Houston, TX 77479.

2Group Leader and Technician, respectively, Intermediates Technology, ExxonMobil Chemical Company, P.O. Box 5200, Baytown, TX 77522.

Emulsifiable concentrates (EC) have been the most common and economical formulation type in the agricultural industry for many years. The basic concept of this formulation involves the dissolution of the active ingredient into a water insoluble inert fluid. Emulsifiers are then added to the solution to enable mixing in the field with water. The choice of EC formulations has become even more popular recently with the rapid development of generic active ingredients. Due to the loss of patent protection and the move to biotechnology-oriented science many new generic active ingredients are being made available to the industry. In many of these cases the formulator is looking for the most economical formulation to enter the marketplace. Other formulation types such as water dispersible granules or liquid suspension concentrates are often considered but require much more intensive processing costs and in some cases lower biological effectiveness than the EC formulation.

When considering EC formulations the formulator must often balance many different attributes of the fluid selected. These attributes may include: flash point, water solubility, packaging compatibility, and reactivity of the fluid. All of these attributes are evaluated once the solubility of the active ingredient in the fluid is measured. Certain standard solubility test methods attempt to categorize these various findings to give the formulator a trend line to the proper fluid. The use of these test methods often will minimize the number of fluids tested and lead to a quicker development timeline.

Many publications have been written on selection of the proper fluid and this paper does not attempt to summarize all the possible methods used for selection of the fluid. In this paper we will discuss the use of three basic solubility tests and the relationships of these tests to actual solubility measurements in the given fluid. The three tests chosen for this evaluation are as follows:

- Test Method for Kauri-Butanol Value of Hydrocarbon Solvents (ASTM D 1133),
- Test Method for Aniline Point and Mixed Aniline Point of Petroleum Products and Hydrocarbon Solvents (ASTM D611), and
- Hildebrand Solubility Parameter (Frisch 1996).

Test results are compared for the different types of fluids. Relationships are then examined between the results of standard solubility test methods (i.e., KB, AP, and Solubility Parameters) and actual solubility of selected generic active ingredients.

Methods and Materials

Fluids and active ingredients were selected which represented the most common materials in the marketplace. In all cases the materials chosen are approved for use under the United States Environmental Protection Agency directive 40CFR180.1001. The fluids selected represent common hydrocarbon and oxygenated solvents and are characterized by distillation range and flash point. They are commercially available from ExxonMobil Chemical Company. Fluids selected are chosen from the following classes: aromatic solvents, paraffinic solvents, and esters (Table 1).

Table 1 - *Properties of fluids studied.*

Fluid	Represented as	Distillation Range, °C	Flash Point, ° C
Normal Paraffin	NP	240-253	105
Mixed Aliphatic 1	MA1	196-214	69
Mixed Aliphatic 2	MA2	224-238	96
Mixed Aliphatic 3	MA3	282-311	141
Isoparaffin	IP	223-254	92
Aromatic 1	A1	185-204	66
Aromatic 2	A2	232-277	93
Aromatic Ester	ARE	256[1]	224
Aliphatic Ester 1	ALE1	176-200	66
Aliphatic Ester 2	ALE2	233-271	118

[1]Represents mid boiling point @ 5 mm Hg.

Active ingredients were selected to represent different types of chemistries with different solubility affinities for the selected fluids (Table 2). All of the active ingredients selected are generic materials technical grade active ingredient which may be purchased from a variety of sources.

Table 2 - *Pesticides used in study (Worthing and Walker 1987).*

Name	Class	Action	Structure
Trifluralin	Dinitroaniline	herbicide	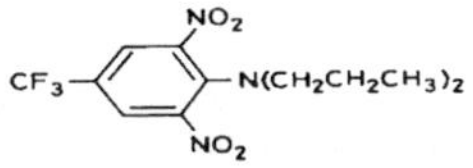
Pendimethalin	Dinitroaniline	herbicide	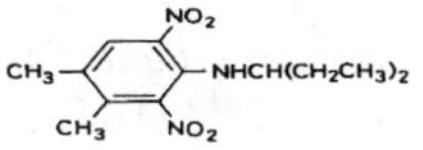
Permethrin	Pyrethroid	insecticide	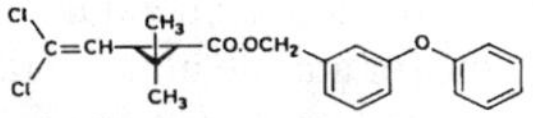
Propanil	Anilide	herbicide	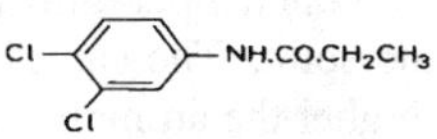
Chlorpyrifos	Organophosphate	insecticide	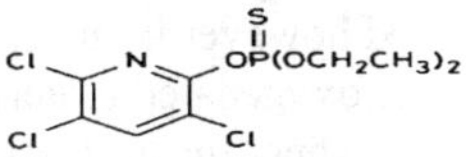
Bromoxynil Octanoate[1]	Phenolic ester	herbicide	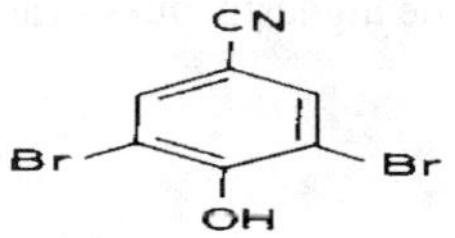

[1]Structure shown for phenolic material not for the ester studied.

Test Methods

Kauri-Butanol Value of Hydrocarbon Solvents (ASTM D 1133) - This method attempts to distinguish between the solubility of a Kauri resin in butanol solution versus the solubility of the resin in a hydrocarbon solvent. As the test material is introduced the turbidity of the resulting mixture is observed. This test has recently become popular as a general solvency method. All types of solvents from hydrocarbon fluids to methyl esters of fatty acids have been characterized by this method. The method also has been effective in determining relative solubility for some esters as well. In this paper we were able to measure values for the hydrocarbon fluids as well as the oxygenated products. High values represent strong solvency. The results of this testing (Table 3) indicate the highest solvency is obtained with the aliphatic esters followed by the aromatic fluids.

Table 3 - *Measured Kauri-Butanol values for test substances.*

Fluid	Observed Value
NP	21
MA2	29
MA1	33
MA3	23
ARE	67
IP	25
A1	93
A2	98
ALE2	101
ALE1	155

Aniline Point and Mixed Aniline Point of Petroleum Products and Hydrocarbon Solvents (ASTM D 611) - This method seeks to relate relative solubility of the fluids by measuring the temperature at which the mixture becomes cloudy. Specified volumes of aniline (or aniline plus heptane in the case of mixed aniline point) and sample are mixed and heated until the two phases are miscible. The mixture is then cooled at a controlled rate and the temperature at which the mixture becomes cloudy (incompatible) is the aniline point. The values obtained indicate the relative aromatic character of the sample. The higher the aromatic character of the sample the lower the value obtained. In most cases the higher aromatic character generally indicates stronger solvency. The test is limited however in its application to this study due to the inability to obtain an endpoint for the oxygenated materials. The findings of the testing would indicate the predicted values where the aromatic materials have the lowest values and the paraffinic materials have the highest values (Table 4).

Table 4 - *Aniline point for test fluids.*

Fluid	Observed Aniline Point
NP	90
IP	90
MA3	89
MA2	79
MA1	69
A1	15
A2	12

Hildebrand Solubility - The determination of the Hildebrand Solubility is based on a mathematical relationship between the molar enthalpy of vaporization and the molar volume (Barton 1991). The relationship can be described as in Equation (1).

$$\delta = ({}_{l}\Delta_{g}H - RT)^{1/2}/V^{1/2} \qquad (1)$$

where

${}_{l}\Delta_{g}H$ = molar enthalpy of vaporization
R = molar gas constant (8.31441 $JK^{-1}mole^{-1}$), and
V = molar volume.

The above mathematical equation establishes a close link between cohesion/energy of vaporization and solubility. Calculation of the Hildebrand Solubility Parameter can be applied for all types of solvents. In the case of this paper many values were determined empirically. Using values determined for pure components of the various products the Hildebrand Solubility Parameter is estimated. A relative relationship can then be expressed between the various materials studied (Table 5).

Table 5 - *Hildebrand Solubility Parameter (δ) for the test fluids.*

Fluid	δ Predicted
NP	7.2
MA1	7.2
IP	7.2
MA2	7.3
MA3	7.4
ALE1	8.2
ALE2	8.3
A1	8.5
ARE	8.6
A2	8.6

Fairly robust consideration of the data would indicate the Hildebrand Solubility Parameter is in close agreement with the previous discussed methods. The trend of the aromatic fluids having the strongest solvency is indicated in this test as well.

Solubility of Active Ingredients - This method involves the actual measurement of the solubility for the individual active ingredient in a given fluid. The procedure involves the addition of active ingredient to the test fluid to a level of a 60% wt./wt. mixture. Solubility is noted and the following action taken in accordance with the observations:

- If crystals of the active ingredient are present, fluid is added to reduce the active ingredient concentration by 5% wt./wt.
- If no crystals are visible and the mixture appears soluble, the solution is seeded with crystals of the active ingredient to check for supersaturation. If there is none, the active ingredient concentration is increased by 5% wt./wt..

The solubility studies continue as above until the maximum solubility is achieved. The test conditions were simulated at 0°C and 25° C. The measured solubility for each active ingredient is represented by plotting the solubility value versus the fluid tested. Figures 1 to 6 display the findings.

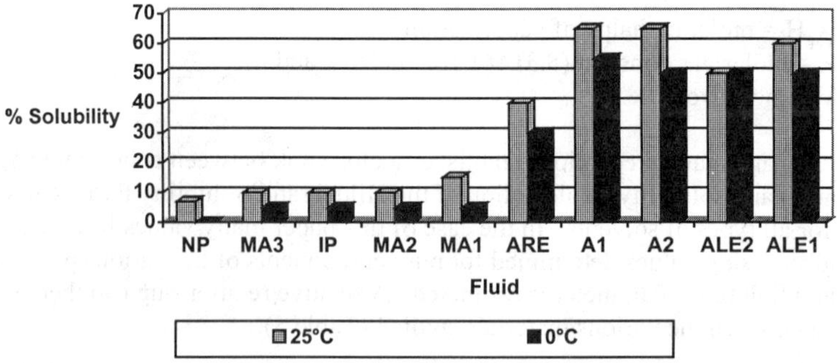

Figure 1 - *Solubility of Trifluralin.*

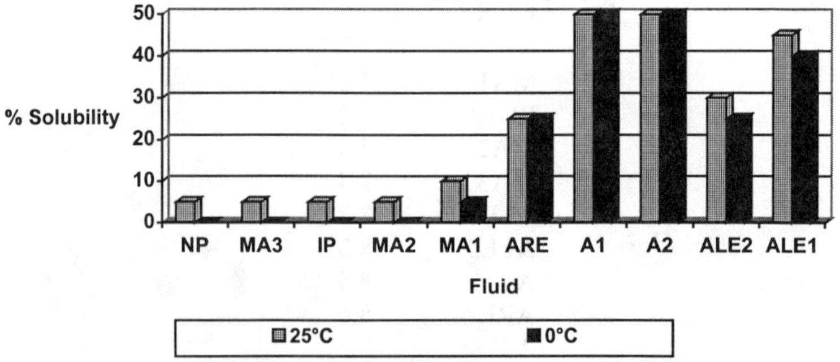

Figure 2 - *Solubility of Pendimethalin.*

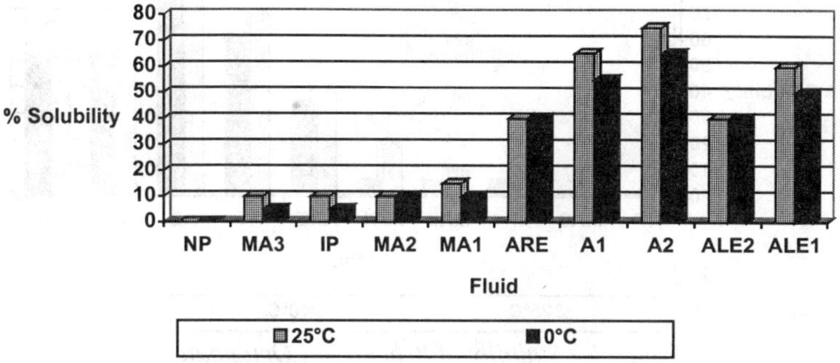

Figure 3 - *Solubility of Permethrin.*

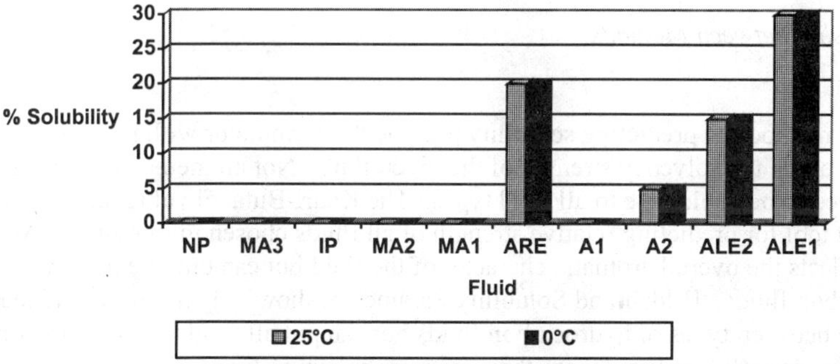

Figure 4 - *Solubility of Propanil.*

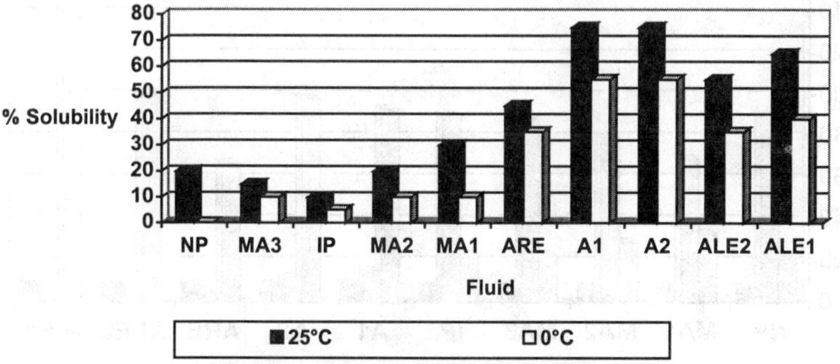

Figure 5 - *Solubility of Chlorpyrifos.*

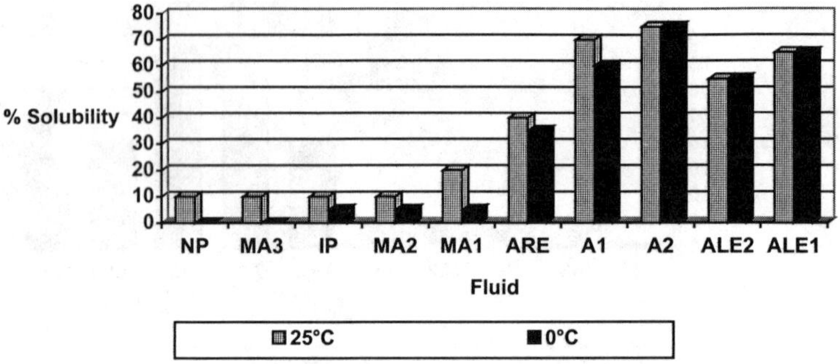

Figure 6 - *Stability of Bromoxynil Octanoate.*

Results and Discussion

Relationship between Methods

All three methods of predicting solubility provide the formulator with relative information on the solvency strength of the given fluid. Not all methods however were determined to be applicable to all fluid types. The Kauri-Butanol value represents an excellent tool for predicting relative strength of all fluids chosen in this study. Aniline Point reflects the overall aromatic character of the fluid but can only be used with hydrocarbon fluids. Hildebrand Solubility Parameters show only minor differences in solvency between types of hydrocarbon fluids but can put all fluid types onto a common scale . The overall trends for these three tests are shown in Figure 7.

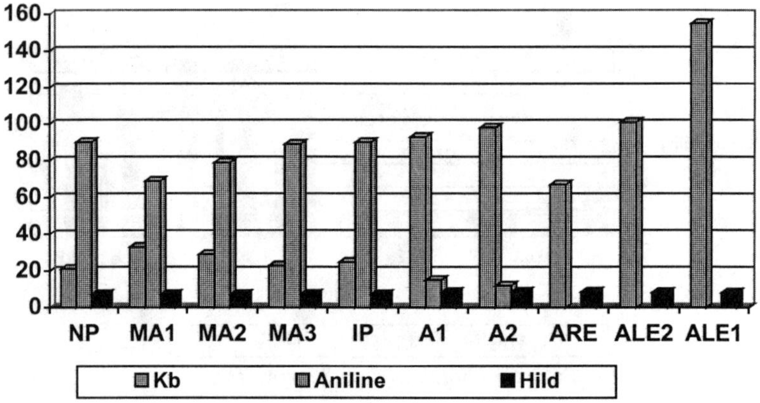

Figure 7 - *Summary of test methods for solvency.*

Relationship of Solubility Methods versus measured Solubility of Active Ingredient(s)

Kauri-Butanol versus Measured Solubility - The Kauri-Butanol Standard Method was developed for and is applicable to hydrocarbon fluids. As such, plotting of the KB value versus the actual solubility information reveals an excellent correlation between these two tests (Figure 8). In all cases the solubility increases with increasing Kauri-Butanol values.

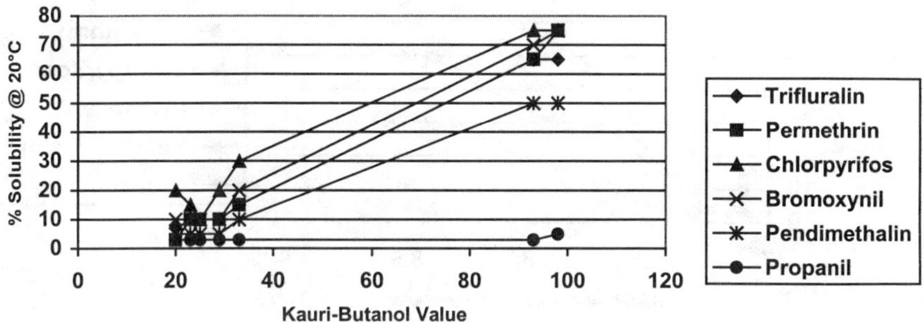

(Values only plotted for hydrocarbon based fluids).
Figure 8 - *Kauri-Butanol value vs. solubility of active ingredient.*

Aniline Point versus Measured Solubility - As in the case of Kauri-Butanol values, the Aniline Point data correlates very nicely with the actual measured solubility. Plotting of the Aniline Point values for the hydrocarbon fluids versus the measured solubility of the active ingredient is displayed in Figure 9. The application of this solvency measurement appears to be a useful tool when applied within a specific fluid family (e.g. hydrocarbon fluids).

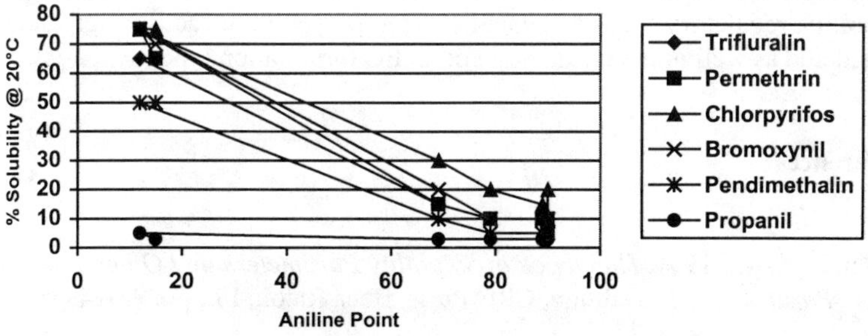

(Values only plotted for hydrocarbon based fluids).
Figure 9 - *Kauri-Butanol value vs. solubility of active ingredient.*

Hildebrand Solubility Parameter versus Measured Solubility - The determination of the Hildebrand Solubility Parameter correlates to the actual solubility data as well as the previous two test methods. The smaller differences observed in these values make the interpretation more difficult but as a general trend there appears to be quite good agreement in the relationship between the solubility and this test method. Once again data is plotted for hydrocarbon fluids only in Figure 10.

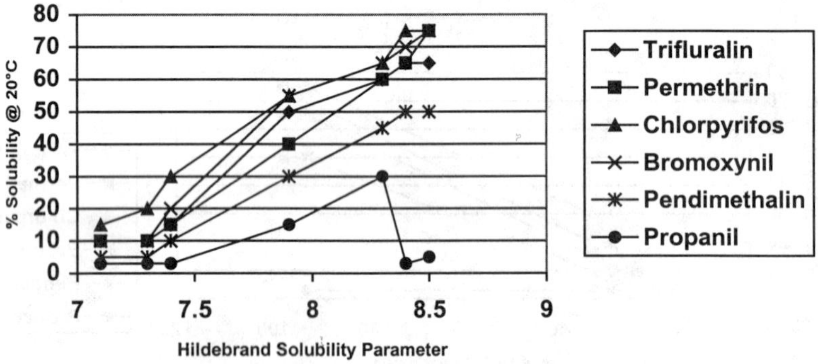

(Values only plotted for hydrocarbon based fluids).
Figure 10 - - *Hildebrand Solubility Parameter vs. Solubility of active ingredient.*

Conclusions

The data above indicates that a linear relationship exists between active ingredient solubility and the three solubility test methods evaluated. In all cases limitations existed when drawing these conclusions to individual fluid families (e.g. hydrocarbon fluids). All properties of the solvent must be considered along with the chemistry of the active ingredient to determine the optimum formulation. In addition to solubility examined in this paper, other properties of a fluid such as: flash point, water solubility, regulatory status and chemical reactivity with the active ingredient need tobe considered as well in the formulation of an EC formulation type.

References

Barton, F. A. M., 1991, *Handbook of Solubility Parameters and Other Cohesive Parameters*, 2nd edition, CRC Press, Boca Raton, FL, pp. 23-24.

Frisch, P. D., "The Application of Solubility Parameters to Agricultural Chemical Problems," Pesticide Formulation and Application Systems: 16th Volume, ASTM

STP 1312, Michael J Hopkinson, Herbert M. Collins, G. Robert Goss, Eds., American Society for Testing and Materials, West Conshohocken, PA, 1996.

Worthing, C. R. and Walker, S. B., 1987, *The Pesticide Manual*, 8th edition, The British Crop Protection Council, Thornton Heath Great Britain.

Jane A. Cooper,[1] Robert D. Fox,[2] Robin D. Dexter,[3] and Martin J. Bukovac[4]

Dynamic Surface Tension of Selected Nonionic Agricultural Surfactants at Surface Ages Relevant to Component Processes in Spray Application

REFERENCE: Cooper, J. A., Fox, R. D., Dexter, R. W. and Bukovac, M. J., "**Dynamic Surface Tension of Selected Nonionic Agricultural Surfactants at Surface Ages Relevant to Component Processes in Spray Application,**" *Pesticide Formulations and Application Systems: A New Century for Agricultural Formulations, Twenty First Volume, ASTM STP 1414*, J. C. Mueninghoff, A. K. Viets, and R. A. Downer, Eds., American Society for Testing and Materials, West Conshohocken, PA, 2001.

Abstract: Dynamic surface tension data for selected nonionic surfactants, pesticides and other miscellaneous liquids are presented. Surface tension was measured using the oscillating jet technique at surface ages ranging from 0.5 to 7 ms. These early measurements are particularly important since new surfaces are being rapidly created, and they represent the time frame of spray droplet formation and impaction processes.

Keywords: Dynamic surface tension, surface age, agricultural surfactants, pesticides, oscillating jet technique, spray application, droplet formation, droplet impaction

The dynamic surface tension (DST) of numerous surfactants has been measured in our laboratory. Some of these data have been reported elsewhere (Cooper and Hall 1993, Brazee et al. 1994, Bechtel et al. 1995, Bukovac et al. 1995, Downer et al. 1995, Reichard et al. 1997, Fox et al. 1998, and Downer et al. 1999), but much of our data have not been published. Since our DST measurements were made at surface ages relevant to important spray application processes, and DST measurements at surface ages less than 10 ms are not generally available in the literature, we provide herein a compilation for a number of nonionic surfactants commonly used in agrochemical formulations and in spray application.

Experimental Methods

Instrumentation and methodology used were as described previously (Reichard et al. 1997), except that the data for Kinetic®, Regulaid®, Triton® X-, and ethyl alcohol were

[1]Research Associate, Entomology Department, Ohio State University/OARDC, 1680 Madison Ave, Wooster, OH 44601.

[2]Agricultural Engineer, USDA-Agricultural Research Service, ATRU, 1680 Madison Ave, Wooster, OH 44691.

[3]Research Fellow, BASF Corporation, Agro Research Division, P. O. Box 400, Princeton, NJ 08543.

[4]Professor Emeritus, Department of Horticulture, Michigan State University, East Lansing, MI 48824.

obtained using a refinement in the lighting procedure. In place of the high-speed stroboscope, a 10 x 15 cm flat fiber-optic light (Lumitex, Inc., 8443 Dow Circle, Strongsville OH) was used. This system produced a light that was uniform in intensity over its entire area. A lower camera magnification (43 X) was used so that the entire fluid waveform, from orifice to jet breakup, could be viewed with one camera placement. This modification both increased the accuracy, and decreased the time taken to measure a test liquid. Data generated using the modified lighting technique are the mean of three replications, other data are not replicated. The materials selected, source and concentrations measured are given (Table 1).

Results

As shown (Table 1), a wide variety of compounds were tested. With the exception of water, ethyl alcohol and isopropyl alcohol, the compounds were not very pure or well defined, although this is the form in which they are extensively used in commercial agriculture. The proprietary materials may also contain active ingredient(s) and additives in varying proportions that may vary from lot-to-lot, year-to-year etc.

DST results are presented (Figures 1 to 16) for the surfactants tested. In most cases, the figures contain one surfactant at several concentrations, or several surfactants with similar chemistry at the same concentration. However, we did not have sufficient samples to make these comparisons for all surfactants. Therefore several unrelated surfactants at selected concentrations are displayed together (Figures 7 and 8).

Nonionic Surfactants

The DST of surfactants of different chemical structure, Triton, Surfynol®, Rhodasurf® DA-, and Igepal® CO- were measured. In these examples the hydrophobic portion of the surfactant remains the same; while the hydrophilic portion is ethoxylated to increasing extents to give rise to molecules of different weights. However, the effect of molecular weight (MW) on DST is not discussed, because along with changes in MW there are also changes in chain length, which in turn affects diffusion and solubility properties, and possibly changes in the additives, or different concentrations of the same additives – so it is likely that more than MW may be different among the surfactants.

Concentration Effect - For all surfactants tested, DST decreased with increased surfactant concentration. This could be due to a simple concentration effect - more surfactant molecules provide for a greater driving force for movement of surfactant molecules to newly created surfaces (or close enough that it can quickly diffuse to the surface), or due to a gradual transition from monomer- to micelle-dominated diffusion (Brazee et al., 1994). In most cases the surfactant concentration had to be greater than 0.1% before the surface tension was decreased significantly from that of water (72 mN m^{-1}), particularly at the youngest surface age measurable.

Table 1-- *Summary of Liquids Tested.*

Material	Chemical composition	Source	Rate(s) tested (%[1])	Figure
1. Nonionic Surfactants				
Igepal® CO-530	Nonylphenoxypoly(ethyleneoxy) ethanol; EO, n=6	ISP Specialties Inc.	0.05	1
Igepal CO-610	EO, n = 9	ISP Specialties Inc.	0.05 & 0.25	1 & 2
Igepal CO-660	EO, n = 10.4	ISP Specialties Inc.	0.05 & 0.25	1 & 2
Igepal CO-720	EO, n = 12	ISP Specialties Inc.	0.05 & 0.25	1 & 2
Igepal CO-730	EO, n = 15	ISP Specialties Inc.	0.25	2
Induce®	P.B.[2] Polyoxyalkylene and free fatty acids	Helena Chemical Co.	0.125	8
Kinetic®	P.B. polyalkyleneoxide modified polydimethylsiloxane and non-ionic surfactants	Helena Chemical Co.	0.125, 0.25, 0.5 & 1.0	3
Regulaid®	Polyoxyethylenepolypropoxypropanol, 2-butoxyethanol, dihydroxy propane	Kalo Ag Chemicals, Inc.	0.05, 0.1, 0.25, 0.5 & 1.0	4
Rhodasurf® DA-530	Polyoxyethylated isodecyl alcohol EO, n = 4	Rhodia Inc.	0.05	5
Rhodasurf DA-630	EO, n = 6	Rhodia Inc.	0.05 & 0.25	5 & 6
Rhodasurf DA-639[3]	EO, n = 6	Rhodia Inc.	0.05 & 0.25	5 & 6
Rhodasurf BC-840	EO, n = 3	Rhodia Inc.	0.5	7
Silwet™ L-77	Polyalkyleneoxide modified heptamethyltrisiloxane	Witco Chemicals, Inc.	0.02, 0.05, 0.1 & 0.25	9
Soprophor® FLK	Ethoxylated polyarylphenol phosphate potasium salt in propylene glycol	Rhodia Inc.	0.05 & 0.25	7
Surfynol® 440				

Table 1-- *Summary of Liquids Tested.*

Material	Chemical composition	Source	Rate(s) tested (%[1])	Figure
Surfynol 465	EO, n = 10	Air Products, Inc.	0.05 & 0.25	10 & 11
Surfynol 485W	EO, n = 30	Air Products, Inc.	0.05 & 0.25	10 & 11
Triton® CS-7	B.: alkylaryl polyethoxylate, sodium salt of alkyl sulfonated alkylate	Rohm & Haas Company	0.125	7
Triton X-45	Octylphenol polyethoxyethanol EO, n = 4	Rohm & Haas Company	0.01, 0.05, 0.1, 0.175, 0.25, 0.375, 0.5 & 1.0	12
Triton X-100	EO, n = 10	Rohm & Haas Company	0.01, 0.05, 0.1, 0.25, 0.5 & 1.0	13
Triton X-405	EO, n = 40	Rohm & Haas Company	0.01, 0.05, 0.1, 0.25, 0.5 & 1.0	14
Tween® 20	Polyoxyethylene (20) sorbitan monolaurate (polysorbate 20)	ICI Americas, Inc	0.05, 0.1, 0.25, 0.5 & 1.0	15
Witconate® AOK	Alpha olefin sulfonate, sodium salt	Witco Chemicals, Inc.	0.05 & 0.25	8
X-77	Alkyl polyoxyethylene, free fatty acids, glycols, isopropanol	Valent USA Corp.	0.02, 0.05, 0.1 & 0.2	16
2. Pesticides				
Ambush® EC	Emulsifiable concentrate formulation of permethrin	ICI Americas, Inc	0.1	17
Ambush WP	Wettable power formulation of permethrin	ICI Americas, Inc	0.2	17
Bacillus thuringiensis	Microbial insecticide	Ecogen, Inc	1.4	18
Benlate®	Benomyl, systemic foliar fungicide	Du Pont Company	0.02	18
Manzate®	Dithiocarbamates, contact fungicide	BASF Ag.	0.24	18
Nova®	Systhane, systemic fungicide	Rohm & Haas Company	0.01	17
Pydrin®	Fenvalerate, insecticide	Du Pont Company	0.06	18

Table 1-- *Summary of Liquids Tested.*

Roundup® Ultra	Glyphosate, herbicide	Monsanto	0.94 & 1.25	17
3. Spray Additives				
Bond®	Synthetic latex, primary aliphatic oxyalkylated alcohol	Loveland Industries, Inc.	1.0	19
Crop Oil	Paraffinic hydrocarbon oil		1.0	19
Nalcotrol®	Polyvinyl polymer	Nalco Chemical Co.	0.055 & 0.125	19
Sunit® II	Methylated seed oil	American Cyanamid Company	0.25 & 1.0	19
4. Solvents				
Ethyl alcohol		Aaper alcohol & chemical Co.	0, 10, 30, 50 & 100	20
Isopropanol		Humco Laboratory	2.5, 5, 30, 60 & 100	21
5. Other Compounds				
Rhodamine®	Fluorescent material	Keystone Analine Corp.	0.29	22
Copper ionized water	Water with about 1 ppm copper ions in suspension			22

[1]Rates tested were measured wt/wt, with the exceptions of the solvents and Rhodamine which were vol/vol.
[2]P.B. – proprietary blend, B. – blend
[3]Rhodasurf 639 is a 90% formulation of Rhodasurf 630.

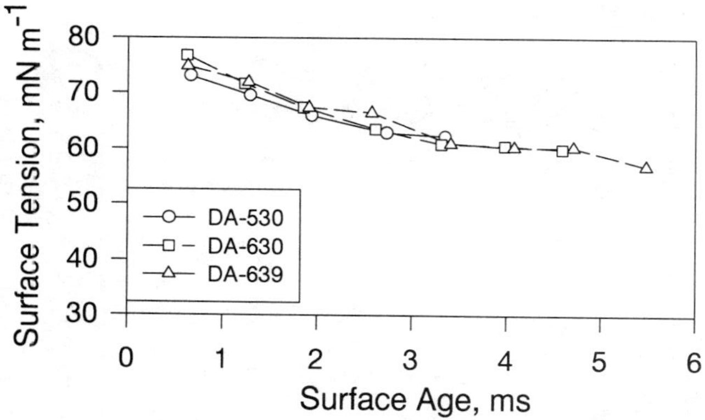

Figure 5 - *Surface tension for 0.05% Rhodasurf.*

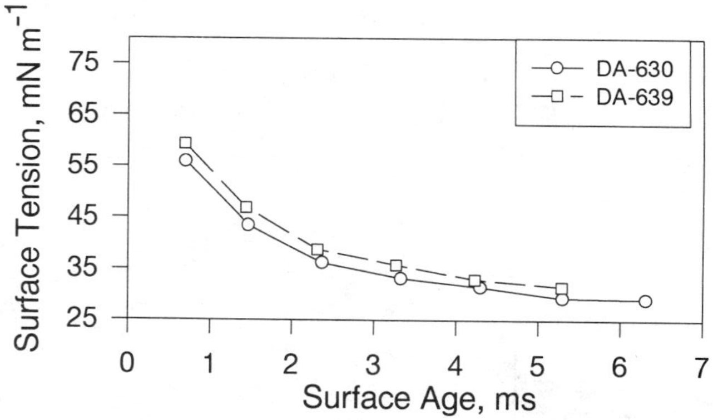

Figure 6 - *Surface tension for 0.25% Rhodasurf.*

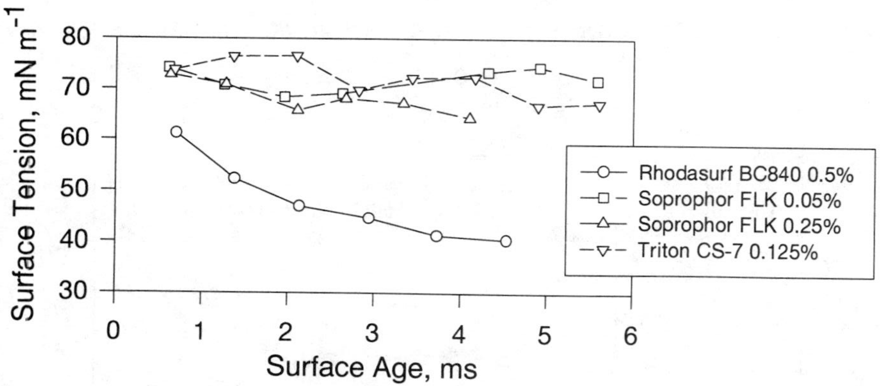

Figure 7 - *Surface tension for Rhodasurf, Soprophor and Triton.*

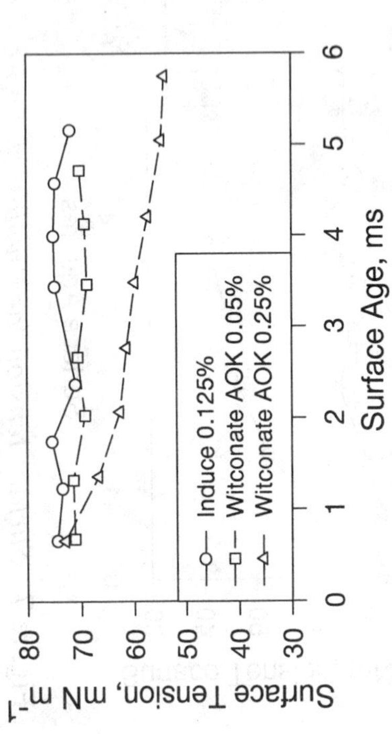

Figure 8 - *Surface tension for Induce and Witconate.*

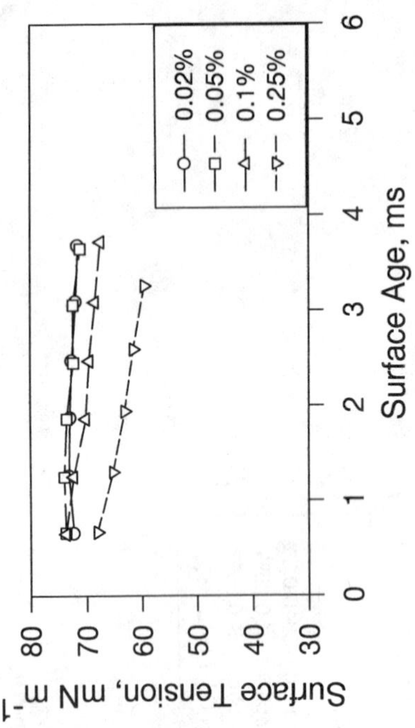

Figure 9 - *Surface tension for Silwet L-77.*

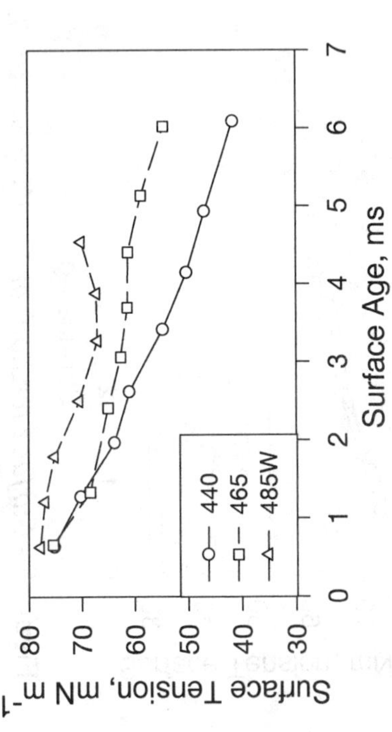

Figure 10 - *Surface tension for 0.05% Surfynol.*

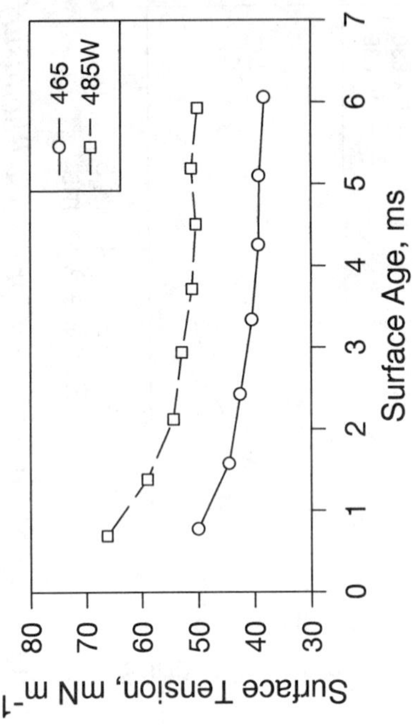

Figure 11 - *Surface tension for 0.25% Surfynol.*

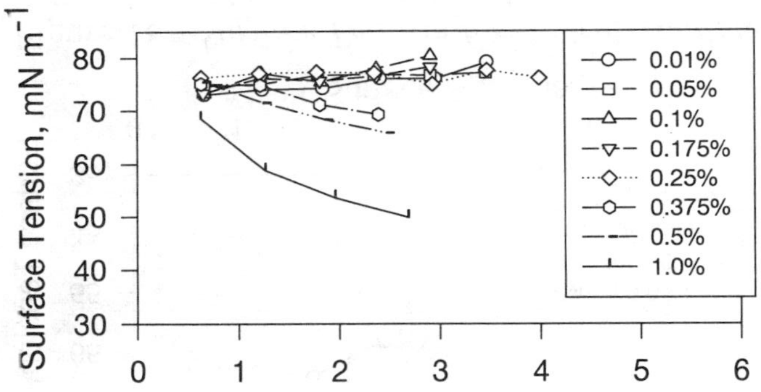

Figure 12 - *Surface tension for Triton X-45.*

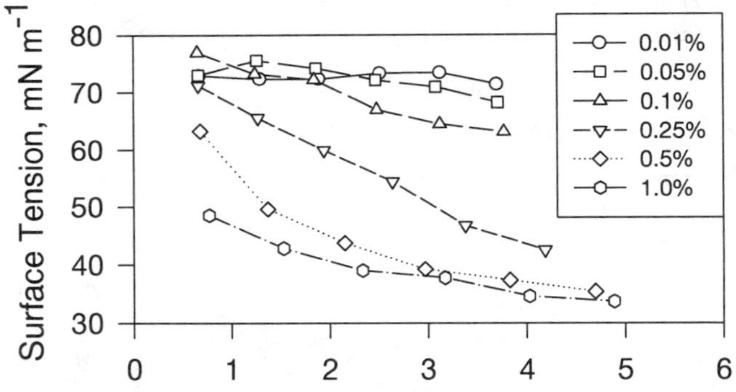

Figure 13 - *Surface tension for Triton X-100.*

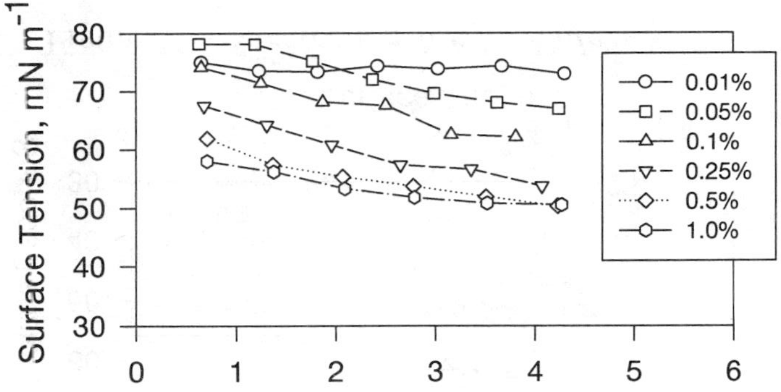

Figure 14 - *Surface tension for Triton X-405.*

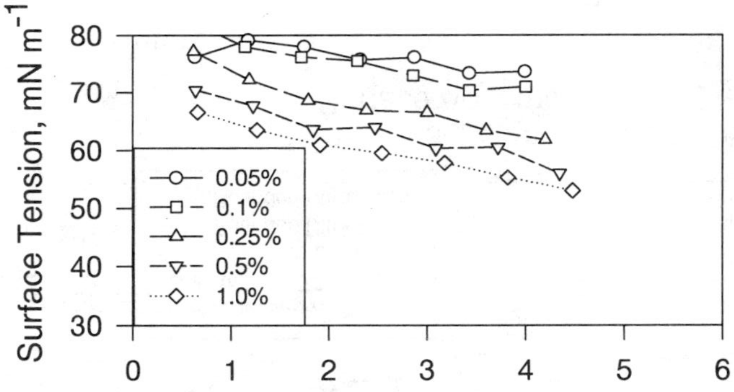

Figure 15 - *Surface tension for Tween 20.*

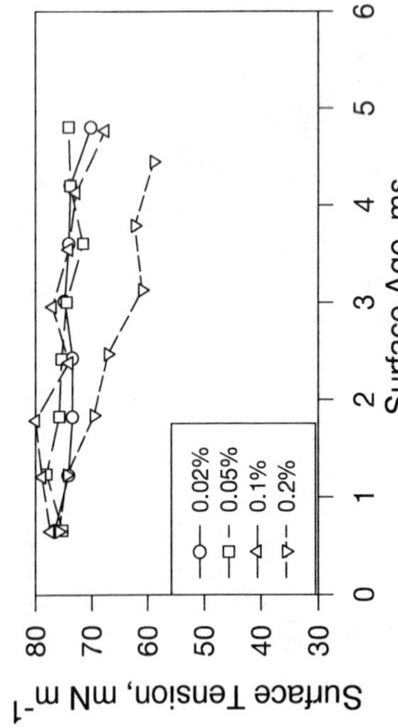

Figure 16 - *Surface tension for Valent X-77.*

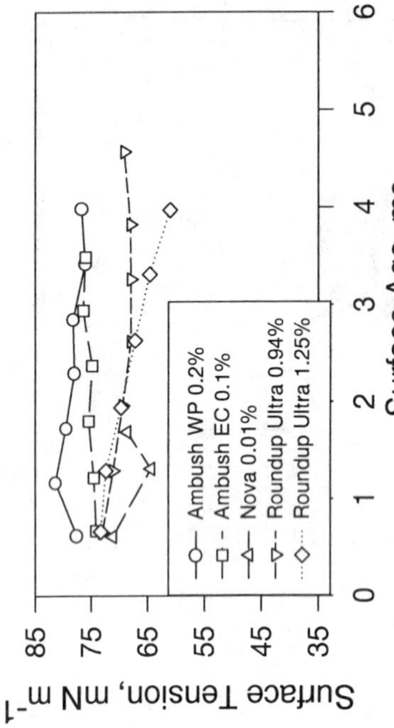

Figure 17 - *Surface tension for Ambush, Nova, and Roundup Ultra.*

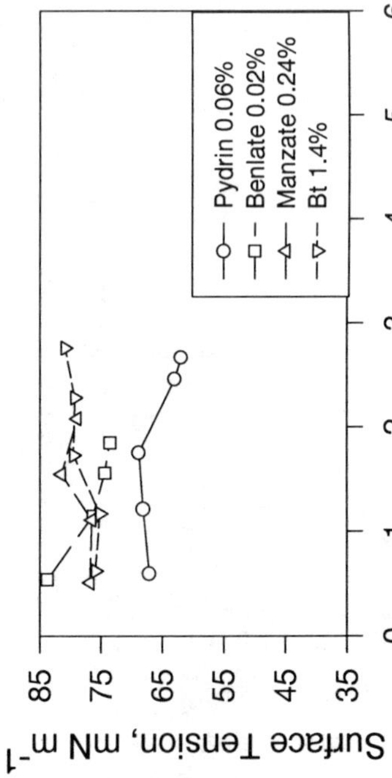

Figure 18 - *Surface Tension for Pydrin, Benlate, Manzate and Bacillus thuringiensis.*

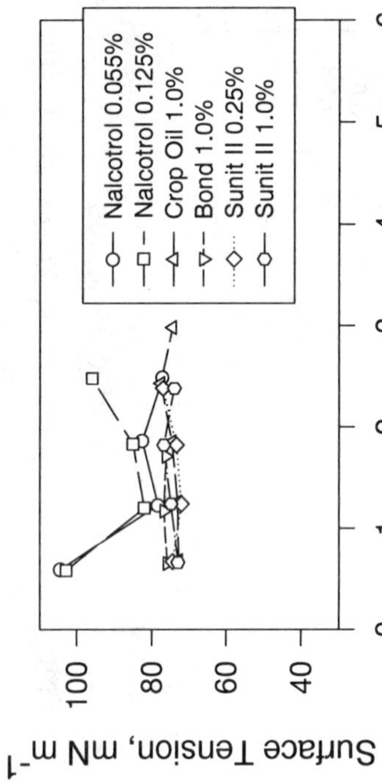

Figure 19 - *Surface tension for Nalcotrol, Crop Oil, Bond and Sunit II.*

Pesticides

None of the pesticides tested (Figures 17 and 18) had DST values greatly different the DST of water. All were tested at label recommended rates.

Spray Additives

Nalcotrol is an agricultural adjuvant, used as a drift retardant, with a relatively high viscosity (7.7 cP unsheared and 3.2 cP @ 20,000 sec^{-1}). DST was greater than water at all ages tested, and especially high (100 mN m^{-1}) at the youngest surface age. Since the oscillating jet system cannot be used for high viscosity materials (the waveform is cylindrical), it is likely that the high values are a reflection of the system's inability to cope with higher viscosity materials rather than a true value. At the concentrations tested, Bond (a synthetic latex sticker), Sunit II (a methylated seed oil) and Crop Oil (an emulsifiable oil used as a herbicide enhancer) all had DST values very similar to that of water (Figure 19).

Solvents

Pure solutions (such as ethyl alcohol and isopropyl alcohol) have a surface tension value that is independent of surface age; hence the equilibrium and dynamic values at any surface age are identical. Any variation from the equilibrium value at any surface age is thus a measure of the accuracy of the system used to measure surface tension. The relatively straight lines and fact that the dynamic values compare favorably with the equilibrium values (given at 0 ms surface age in Figures 20 and 21), suggest the oscillating jet technique gave an accurate measurement of DST.

Other Compounds

Rhodamine, a water-soluble tracer, and copper ionized water (Figure 22) both had DST values very similar to that of water.

Discussion and Conclusions

The oldest age at which DST could be measured varied with the liquid tested. The waveforms of some liquids were very stable, while others became distorted soon after leaving the orifice. Similarly, different formulations of the same surfactant varied in their stability. In general, DST decreased with both increasing surface age and increasing concentration. At concentrations less than 0.1%, DST of most liquids was nearly equal to that of water.

The method used to determine surface tension was accurate - as indexed by the small deviation of the DST values from the equilibrium values of the pure solutions ethyl and isopropyl alcohol.

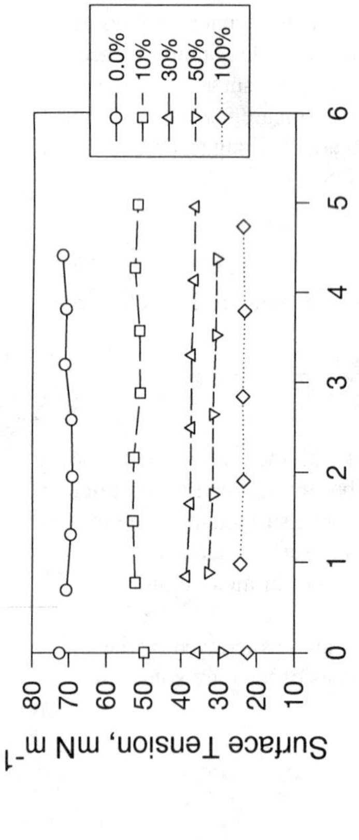

Figure 20 - *Surface tension for Ethyl Alcohol.*

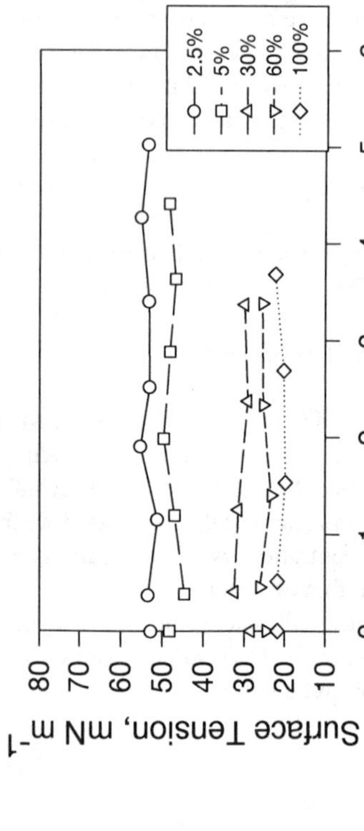

Figure 21 - *Surface tension for Isopropanol.*

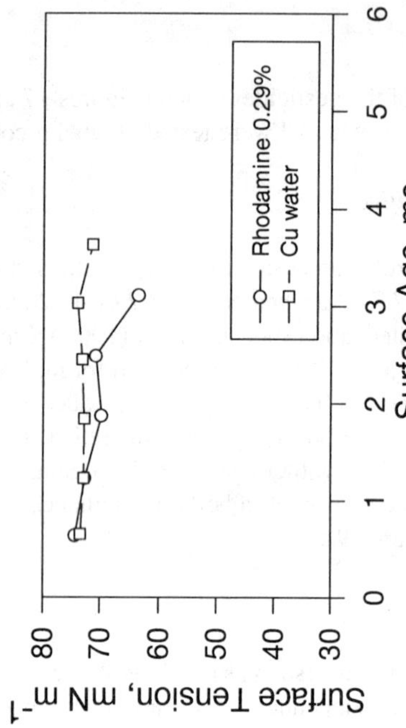

Figure 22 - *Surface tension for Rhodamine and copper ionized water.*

SESSION III: Formulation Strategy

Kozo Tsuji [1]

Recent Trends in Pesticide Formulation in Japan

Reference: Tsuji, K., " **Recent Trends in Pesticide Formulation in Japan,"** *Pesticide Formulations and Application Systems: A New Century of Agricultural Formulations: Twenty First Volume, ASTM STP 1414,* J. C. Mueninghoff, A. K. Viets, and R. A. Downer, Eds., American Society for Testing and Materials, West Conshohocken, PA, 2001.

Abstract: Recently in the pesticide industry in Japan, it has become considerably more desirable to develop products that are safer both for workers and the environment. Moreover, number of farm workers of advanced age and women have gradually increased. Therefore, it is also desirable to develop more efficient and more labor-saving formulations and application technologies in order to make agricultural work lighter. For these purposes, various promising formulations have been developed such as flowable formulations for direct application, water dispersible granules, microcapsules, jumbo herbicides, 1 kg granules, water surface spreading formulations, seedling box treatment, fertilizers containing pesticides, and physical controlling methods. These formulations are reviewed, and their advantages and trends are discussed.

Keywords: flowable formulation, jumbo herbicide, 1 kg granule, water surface spreading formulation, microcapsule, water dispersible granule, fertilizer containing pesticides, physical controlling method, safety, environment, labor-saving

Introduction

Recent requirements of the pesticide industry in Japan are as follows (Tsuji 1998a, 1998b),

(1) Higher safety: Pesticide should be safe for workers and have no harmful effect or impact on nontarget organisms or the environment.

[1] Director, Formulation Technology Research Laboratory, 3-10-9, Shigigaoka, Sango-cho, Ikoma-gun, Nara 636-0813, Japan.

(2) High efficacy: Pesticides should have good initial and residual efficacy at lower dosage.
(3) Lower price: Manufacturing cost should be low and cost performance should be good.
(4) Labor-saving: The application should be easy and application efficiency should be high.

It is not easy for candidate compounds by themselves to satisfy all these requirements. It has thus become very difficult to develop new pesticides, and the cost and time for the development of a new pesticide has increased significantly. Therefore, it becomes important to improve formulations and application technology in order to satisfy the above requirements for new and existing active ingredients. Thus the research work on formulations and application technology has also become very important.

Various improvements in formulations and application technologies have been made according to the recent demands by the pesticide industry. Changes in formulation types have also proceeded and various new types of formulations and application methods have been developed as shown in Tables 1 and 2 (Tsuji 1996, 1997a).

Directions of improvement are as follows;

(1) Using water instead of organic solvents resulting in water-based formulations;
(2) Granulation or water soluble packaging of wettable powder;
(3) Development of microcapsule as a typical controlled release formulation;
(4) Development of various functional formulations such as jumbo herbicides, water surface spreading formulations, and fertilizers containing pesticides;
(5) Development of various physical controlling methods; and
(6) Development of various application methods such as application from levee, application to seedling box, and application at the same time as transplantation.

According to these trends, various kinds of new formulations and application technologies are developed. Water-based formulations, water dispersible granules, water-soluble package, microcapsules, jumbo herbicides, water surface spreading formulations, 1 kg granules, seedling box treatment, fertilizers containing pesticides and physical controlling methods will be explained in more detail.

Water-Based Formulations

Water-based formulations (Tsuji 1994, 1996, 1997a, 1998a, 1998b) are obtained by emulsifying or suspending pesticide technical grade active ingredients in water. These formulations are known under emulsion, oil in water, microemulsion, suspension concentrate, suspoemulsion and

Table 1- *Development of new pesticide formulations*

Formulations	Problems	Reasons	Methods of improvement	New formulations
Emulsifiable Concentrate	Toxicity Phytotoxicity Flammable	Organic solvent Synthetic surface active agent	Making use of water Solidifying Changing solvents or emulsifiers	Emulsion,oil in water (EW) Microemulsion (ME) Supspemulsion (SE) Emulsifiable granule or powder Improved recipe
Wettable powder	Dusty	Fine powder	Dispersing in water Granulating Packaging with water soluble film	Suspension Concentrate (SC,Flowable) Water dispersible granule (WG) Dry flowable Water soluble package
Dust	Drift	Fine powder	Removing very fine particles Granulating	DL dust Fine granule F
Flowable	Deterioration of physical properties	Dispersion in liquid	Removing liquid Improving recipe	Dry flowable Improved recipe
Granule	Weight	3 kg/0.1ha	Higher concentration	1 kg granule

Table 2- *Recent requirements in pesticide industry, methods of improvement, and new types of formulations.*

Recent requirements	Methods of improvement	New types of formulations
	Controlled release	Controlled release formulation (microcapsule, laminate, granule)
	Exposure reduction	Water soluble package Water dispersible granule
	Application from levee	Jumbo herbicide Packed formulation
Improved safety	Application from water inlet of paddy field	Water surface spreading formulation
Labour-saving	Application of pesticides to seedling box	Seedling box formulation
Improved biological efficacy	Application of pesticides at the same time as transplantation	
	Seed treatment	Formulation for seed treatment
	Application of pesticides with fertilizers	Fertilizer containing pesticide
	Making use of biopesticide	Biopesticide formulation
	Targeting	Water surface spreading formulation
	Physical controlling method	Sticky formulation

multiple emulsion.

Water is the safest liquid. Therefore, water-based formulations have various advantages from a safety point of view as follows:

(1) Reduced toxicity and irritation,
(2) No dust problems,
(3) Nonflammable,
(4) Better efficacy than WP due to smaller particle size,
(5) Reduced phytotoxicity,
(6) Nonstaining,
(7) Reduced smell,
(8) Plastic bottles can be used.

Therefore the water-based formulations with these superior advantages have been developed and commercialized for many pesticides.

In Japan herbicide flowable formulations have been developed for direct application for labor-saving, i.e. without dilution (Takeshita 1994). These flowable formulations have good diffusion properties in water and can be applied without dilution by hand from levees of paddy fields. Application rate is 500~1000 mL/10a. In this case, spreadability of the formulation in water is improved by smaller particle size and /or selection of suitable surface active agents. Therefore pesticides are spread uniformly in the paddy field.

The concentration of the active ingredients (AI) reaches an even level within 1 day in paddy water and within 3 days on soil surface. In this case, water level is kept 4~5 cm high.

In order not to stick to rice and to reduce phytotoxicity, surface tension is adjusted to be 36~65 dyne/cm at 25C (Takematsu et al 1995).

In the case of application from the levee, time for application is reduced to one fifth of herbicide granule application in the paddy field.

For labor-saving purposes, the application of a herbicide flowable formulation to water irrigation inlet of the paddy field is also carried out, when water is let into the paddy field immediately after transplantation. In this case, 500~1000 mL of flowable formulation is applied and it spreads over the paddy field by water flow.

Water Dispersible Granules

Water dispersible granules (WG) (Imai et al 1990, Tsuji 1994, 1996, 1997a, 1998a, 1998b) were developed to prevent dustiness normally associated with wettable powder formulations at the time of dilution.

This formulation disintegrates and disperses readily in water after addition.

The advantages of this formulation are as follows:

(1) Dust free and safe for workers,
(2) High density and compact packaging,
(3) Constant apparent density and measurable by volume,

(4) Good flowability and superior handling,
(5) Possibility of high concentration formulation,
(6) Packageable in paper bags and various containers, and
(7) Little residue in the container and the used containers are easily disposed.

WG can be formulated by various methods. Recently in Japan, the extrusion method has become more popular.

Some WGs are packed in water–soluble packages.

Water Soluble Packages

Water soluble packages (Suzuki 1996) of pesticides have various advantages as follows:

(1) Reduction of worker exposure at the time of dilution of WP,
(2) Safe and easy handling,
(3) Unit dose convenience, and
(4) Uncontaminated package waste.

Polyvinyl alcohol is mainly used as a water-soluble film.
The physicochemical properties of PVA have been improved significantly, including water solubility, physical strength, processability, and stability.

In Japan, herbicide granules packed by water-soluble film have been developed for labor-saving applications. This formulation is one kind of jumbo herbicide which are described later. This packaged formulation is thrown by hand from levees into a paddy field. Recently, a similar application method of granules packaged by water-soluble film has been developed for insecticides and fungicides. In these cases, most of the granules are spread on water surface to get better biological efficacy.

Water soluble bottles made of PVA are also used for an oil formulation for rice pest control, which spreads over the entire water surface of the paddy field in a few minutes. Six to ten bottles are thrown by hand from the levees into a 0.1 ha paddy field.

Microcapsules

Microcapsules (MCs) (Ohtsubo and Tsuji 1997, Tsuji 1990, 1993, 1996, 1998a, 1998b, 1999a, 1999b) are small particles(1~1000 microns) composed of a core material and an outer wall. The wall isolates the core material from the environment and protects it from environmental degradation and interaction with other materials.

The general advantages of pesticide MCs are as follows:

(1) Controlled or slow release of core AI to result in improvement of residual activity,
(2) Longer application interval resulting in labor-saving,

(3) Reduction of application dosage,
(4) Stabilization of core AI against environmental degradation,
(5) Reduction of mammalian toxicity,
(6) Reduction of human mucous-membrane irritation,
(7) Reduction of phytotoxicity,
(8) Reduction of fish toxicity,
(9) Reduction of evaporation and leaching,
(10) Reduction of environmental pollution,
(11) Reduction of reactivity of two pesticides,
(12) Masking of odor,
(13) Solidification of liquid pesticides,
(14) Reduction of drift,
(15) Increase in the number of target organisms,
(16) Consistent activity irrespective of application surfaces, and
(17) Easier handling.

The MC has several advantages, which meet the current demand of pesticide industry. Therefore the research and development of pesticide MC has grown significantly in recent years especially in the agricultural field, although the initial step of the development was mainly for household uses.

Examples of commercial pesticide MCs available in Japan are summarized in Table 3 (Tsuji 1999b). Some examples are explained.

Fenitrothion MCs for cockroach control were developed (Tsuji 1990, 1998a, 1999a, Tsuji et al. 1989). The wall material is polyurethane. They were prepared by interfacial polymerization. These MCs at the rate of 125 mg of AI/m^2 caused 100 % mortality to the German cockroaches even 8 week after treatment (Kawada et al. 1990). Such a long-lasting residual activity is caused by the trampling mechanism, that is, the cockroaches broke the MCs when they contacted the MCs (Tsuda et al. 1987, Tsuji et al. 1989).

If trampling is the mode of action, biological efficacy is dependent on the strength of the MC. It was found that the trampled percentage of MC by the contact of cockroaches can be controlled by D/T as shown in Figure 1, where D is a mass median diameter and T is a wall thickness (Ohtsubo et al. 1987, Tsuji et al. 1989). D/T is known to be a parameter of MC strength. When the value of D/T is set properly, both initial and residual efficacy were good. Discussion of physical strength of the MC was also reported (Ohtsubo et al. 1991).

Fenitrothion MC was developed for termite control (Ito et al. 1988, Tsuji 1990, 1999a, Tsuji et al. 1989). The MC has high biological efficacy and quite long residual activity against the termite, Coptotermes formosanus Shiraki, both on wood and in soil. Fenitrothion is released from the MC only by breaking of the MC, not by diffusion.

Table 3- *Examples of commercial pesticide microcapsules available in Japan.*

Product	Active Ingredient	Conc.(%)	Main Target
Sumithion MC	Fenitrothion	20	Bugs(paddy field, Aerial application)
Sumithion MC	Fenitrothion	5	Cockroach(bait)
Sumithion MC for cockroach control	Fenitrothion	20	Cockroach (residual spray)
Kareit MC	Fenitorthion	20	Termite
Lumbert MC	Fenitrothion	20	Powder pest beetle (plywood pannel)
Sumipine MC	Fenitrothion	23.5	Japanese pine sawyer (forest, aerial application)
Sumibassa MC	fenitrothion/ fenobcarb	10/15	Bugs, planthopper (paddy field, aerial application)
Sumi Cue-lure Microcapsule Sol	Fenitrothion/ Cue-lure	10/3	Melon fly
Diazinon SL Sol	Diazinon	25	Gold beetle
Diazinon MC Suspension	Diazinon	24	Cockroach (residual spray)
Kayatack MC	Chlorpyrifos	25	Termite
Lentrek 20MC	Chlorpyrifos	20	Termite
Baktop MC	Fenobcarb	15	Termite
Ember MC	Permethrin	10	Rice water weevil (seedling box)
Aniverse MC	Halfenprox	5	Mites
Trebon MC	Ethofenprox	20	Planthopper, bugs, leefhopper,(paddy field, ground application)
Trebon Sky MC	Ethofenprox	20	Planthopper, bugs (paddy field, aerial application)
Mocap 3MC	Ethoprophos	3	Nematode
Guardjet WP	BT	7	Cabbage moth, larva of small white butterfly
Oncol MC	Benfracarb	20	Cabbage moth
Ohte FL	Esprocarb(MC)/ Bensulfuron methyl(FL)	30/1.4	Barnyard grass
DEET MC	DEET	36	House dust mite
Naramycin D80	Cycloheximide	80	Rat
Ratden S	Capsaicins	2	Rat

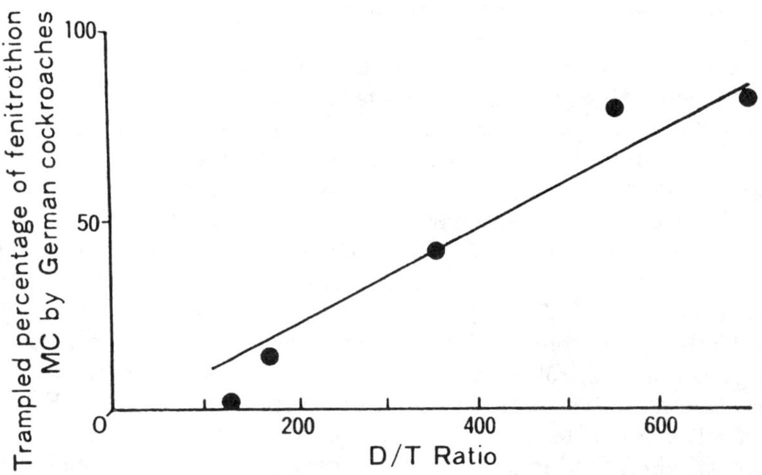

Figure 1- *Relationship between the percentage of fenitrothion MCs trampled by German cockroaches on a petri dish and the D/T ratio.*

Aerial concentration of fenitrothion near the operator's mouth during spraying for 20~40 min under the floor was 0.07~0.4 mg/m^3, which is low compared with the concentration allowable in the work environment (1 mg/m^3). Adhesion of fenitrothion to the operator's clothing and skin was also very low (Tsuji et al. 1989).

Fenobcarb MC (Baktop MC) also shows both good initial and residual efficacy to termite (Ohtsubo et al. 1996). This MC is supposed to act as stomach poison.

Fenitrothion MC (Lumbert MC) was developed for insect-proof plywood panels in order to improve chemical stability of fenitrothion in the alkaline resin (Kawashima et al. 1991, Tsuji 1990, Tsuji et al. 1989).

The MC is used by mixing with the glue for making the plywood panel. Plywood containing fenitrothion MC was completely protected from boring by *Lyctus brunneus* , even in the condition where plywood containing fenitrothion EC was penetrated. The mode of action for *Lyctus brunneus* is hypothesized to be breaking of the MC by biting and contact with fenitrothion.

There are reported fenitrothion and pyrethroid MCs for agricultural use whose mode of action is breaking of MC (Tsuji 1990, 1998a, 1998b, 1999a, 1999b). As for the biological efficacy of fenvalerate MC against the diamondback moth, a linear relationship was observed between LC_{50} and D/T ratio, a parameter of the strength of MC, on a log-log graph paper (Ohtsubo et al. 1989).

The phytotoxicity of fenitrothion MC on Chinese cabbage and rainfastness were found to depend on only the wall thickness.

Fish toxicity against killifish of fenvalerate MC was found to depend on DxT as shown in Figure 2 (Ohtsubo et al. 1989, Tsuji 1990). The parameter DxT relates to the release rate of fenvalerate into water from MC. Therefore fish toxicity is controlled by diffusion of AI into water.

Mammalian toxicity LD_{50} (mg/kg) of fenpropathrin MC was also found to depend on DxT (Ohtsubo et al. 1991, Tsuji 1990). From these results, D/T, DxT, and T are said to be good parameters for design of the MC.

Permethrin microcapsule (Ember MC) was developed for rice water weevil control by the seedling box treatment. LC_{50} depended on D/T. Fish toxicity depended on DxT.

Fenitrothion and fenitrothion/fenobcarb MCs were developed for bug and planthopper control in rice paddies by aerial application. MC gave longer residual activity against small brown planthopper than EC. The mode of action seems to be breaking of MC by the insects.

Fenitrothion MC (Sumipine MC) was developed for forest application. This MC was very effective against Japanese pine sawyer, a vector of the pinewood nematode, in comparison with EC, and gave a better withering prevention effect against pine tree than EC. The mode of action is destruction of MC by biting.

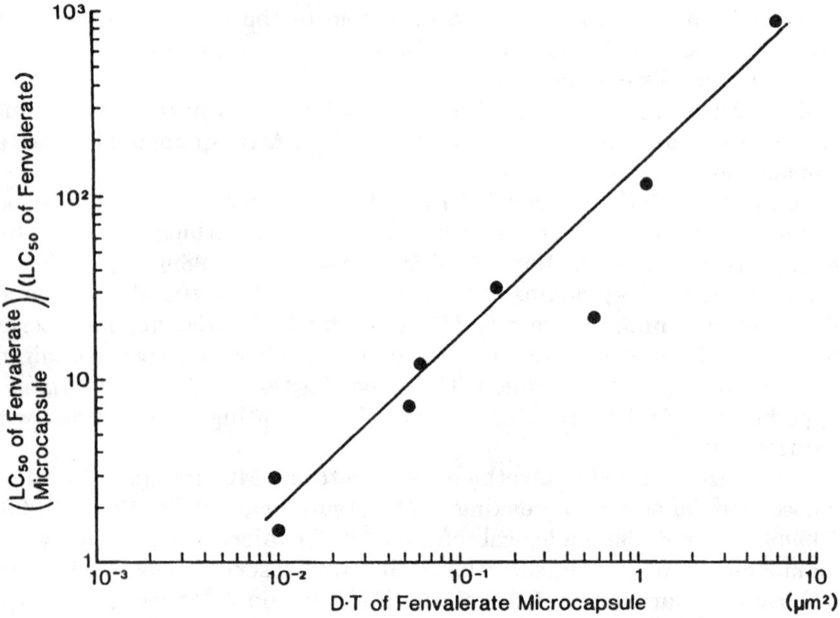

Figure 2- *Fish toxicity of fenvalerate microcapsule to killifish.*

Jumbo Herbicides

Jumbo herbicides (Noritake 1993, Tsuji 1998a, 1998b, Ueda et al. 1997) are throw-in type herbicide formulations, and each formulation is about 50 g and applied at 100~200 units per hectare from the levees by hand. Then they spread and diffuse easily in water.

The advantages of jumbo herbicides are as follows:

(1) Small and light: Dosage is 1/3 of usual application.
(2) Easy application and labor-saving: They can be applied by hand from the levees.
(3) Dust free and no drift: Safe for both workers and adjacent crops.
(4) Proper application: Depending on the area of the field, appropriate number of the formulations are thrown at equal distance.
(5) Reduced cost of rice crop: Costs of delivery and storage of the formulation becomes cheaper.
(6) Applicable even in bad conditions: They can be applied even on windy days.

Jumbo herbicides are classified into two types. One is an effervescent tablet type. The other is water soluble package type. Effervescent tablets are formulated by using solid acid, carbonate and dispersants. When these tablets are put into water, they effervesce vigorously and spread over a broad area. AIs spread smoothly in the paddy field and become uniform within 6~24 hr.

Water-soluble bags of herbicide granules are also available.

Water Surface Spreading Formulation

Water surface spreading granules (Sekiguchi et al. 1991, Tsuji 1998a, 1998b) are formulated by using potassium chloride as a carrier. This formulation is made mainly by impregnation of AI solution into granulated carriers. After application to the paddy field, granules sink temporarily down onto the soil. Then potassium chloride dissolves and granules resurface and float. Finally, the AI solution spreads in all directions on the water surface. This behavior is shown in Figure 3 for cycloprothrin formulation, which was developed for the first time in order to control the rice water weevil (Sekiguchi et al. 1991). Rice water weevils live in paddy water and on rice plants, and touch the water surface. Therefore this formulation can deliver the high concentration of the AIs to the place where target rice water weevils are living, to result in good control. Therefore dosage of AI can be reduced by targeting.

Binders in the granules are very important to resurface the granules. The binders act to trap air in the granules after the dissolution of potassium chloride in water that result in resurfacing of the granules. A

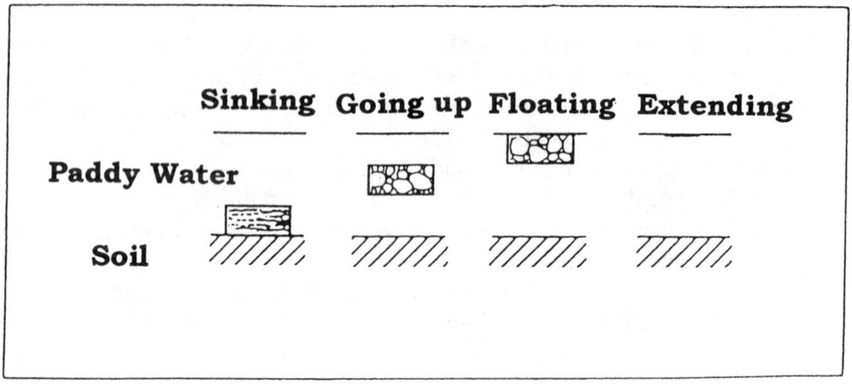

Figure 3- *Behavior of cycloprothrin 2% resurfacing granules in water.*

combination of high molecular weight sodium polyacrylate and xanthan gum is the best.

These granules are also packaged in water-soluble bags, and are applied by throwing ten 60 g bags from levees by hand (Masui 1993). The advantages obtained are similar to those from jumbo herbicides.

This technology has been applied to the herbicides including solid technical materials. This is one type of jumbo herbicide.

Floating carriers such as foaming perlite, wood powder, and silicone surfactants are also applied to help the granules float, spread and diffuse on the water surface after application of water-soluble bags.

A water surface spreading oil formulation has also been developed (Enomoto 1993). This formulation is put drop by drop into the paddy field from bottles, which are hung from poles. Oily AIs immediately spread on the water surface. The oily film adheres to leaf sheaths and leaves of the rice plant. This application method is effective even in rainy condition and it provides for substantial reduction in labor requirement.

1 kg Granules

In Japan, herbicide granules are packed in 3 kg packages that will treat 0.1 hectare. The Japan Association for Advancement of Phyto-Regulators (JAPR) proposed to reduce the dosage to 1 kg/0.1 ha , which is called 1 kg granule (Takeshita 1994, Tsuji 1994, 1996, 1997a, 1998a, Tsuji and Ohtsubo 1993).

The 1 kg granule was developed to save application labor in paddy field and also to reduce distribution and storage charges. The 1 kg granules contain three times the amount of AI as compared to the 3 kg granules. The 1 kg granules are designed to be spread from the levees of the paddy field, which has a 30 m total width. Therefore the particle diameter needs to be 1.0~1.5 mm (mainly 1.2 mm) to reach out 15 m from each levee by the power spreader. Therefore the number of particles of 1 kg granules per unit area is about one eighth of that of 3 kg granules. In order to get consistent herbicidal efficacy, it is essential to maximize the diffusion of the AIs in paddy water by either making smaller particle sizes of the AIs, and/or selection of appropriate surfactants. With recent improvement, the 1 kg granules now give acceptable herbicidal efficacy.

Seedling Box Treatment

Many granules for seedling box treatment are being developed to save labor. In this case, prevention of phytotoxicity and longer residual activity is important. Thus the AI should be controlled released from the granules for seedling box treatment. Therefore the granules are usually coated by selected polymers. This is a MC of the granule, and application dosage can be reduced. Moreover there is no drift and less influence to the environment. Due to these superior advantages, the seedling box treatment is likely to become more common.

Fertilizers Containing Pesticides

Fertilizers containing pesticides (Tsuji 1998b) have been developed as a high performance fertilizers. They have the following advantages which meet current pesticide industry.

(1) Labor-saving: Pesticides are applied at the same time when fertilizers are applied.
(2) Uniform spraying: Pesticides are diluted with large amount of fertilizer. Therefore spraying becomes uniform.
(3)Reduction of phytotoxicity: They do not stick to the plant surface.
(4)Safe handling: Workers do not contact a high concentration of pesticides.
(5)No drift.
(6)Reduction of auxiliary materials of pesticide formulation: Fertilizers become carriers of the pesticides.
(7)Controlled release of pesticides: Pesticides are released slowly.
(8)Efficient absorption of pesticides: Plant roots reach out toward the fertilizers.

Fertilizers containing pesticides are made by a kneading method, coating method, or impregnation method. When proper application timing of pesticides is different from that of the fertilizer, the release rate of pesticides should be controlled.

As fertilizers containing pesticides have various advantages mentioned above, development will proceed.

Physical Controlling Methods

Various kinds of physical controlling methods (Tsuji 1998b) have been developed in Japan, where sodium oleate, potassium oleate, and a mixture of soybean lecithin and purified machine oil and starch are used, for examples. The major advantage of physical controlling methods is to avoid development of pest resistance to chemical treatments.

Starch is food and a very safe material both for workers and to the environment. Starch solution is sprayed to the leaves of the plant, and starch traps mites with its stickiness after drying.

Conclusions and Future Trends

In order to meet recent demands of pesticide industry, various kinds of new formulations and application technologies have been developed. For this purpose, the concept of "Pesticide Delivery System"(PDS) is very important. PDS can be defined as a technique or a method in which active ingredients are made available to a specified target at a concentration and duration designed to accomplish an intended effect, i.e., obtain the fullest biological efficacy while minimizing various harmful effects. Therefore research work along this line should be continued, and targeting and controlled release technologies become very important. The trend is to develop more various stimuli-responsive formulations and application technologies.

It is desirable to develop intelligent pesticide formulations and application technologies, that have sensing, processing, and actuating functions. Then pesticide application will become very efficient, safe and labor saving (Tsuji 1997b).

Reference.

Enomoto, Y., 1993, "Trebon Surf, New Techniques for Rice Pest Control,", *Agrochemicals Japan*, No. 63, pp.4.

Imai, M., Horide, F., Ohtsubo, T., Tsuda, S., and Tsuji, K., 1990, "New Pesticide Formulations," *Sumitomo Kagaku*, Vol. 1990-II, pp. 73-79.

Ito, T., Kohno, Y., Ohtsubo, T., Tsuda, S., Tsuji, K., and Shinjo, G., 1988, "Efficacy of the Fenitrothion Microcapsule against the Termite, *Coptotermes formosanus* SHIRAKI," *Japanese Journal of Applied Entomology and Zoology*, Vol. 33, pp. 301-307.

Kawada, H., Makita, M., Tsuda, S., Ohtsubo, T., Shinjo, G., and Tsuji, K., 1990, "Efficacy and the Way of Action of Fenitrothion Microcapsules for Cockroach Control," *Japanese Journal of Environmental Entomology and Zoology*, Vol. 2, No. 1, pp. 6-13.

Kawashima, M., Ohtsubo, T., Tsuda, S., Itoh, T., and Tsuji, K., 1991, "Application of Fenitrothion Microcapsule for Insect-Proof Plywood,"

Journal of Pesticide Science,, Vol. 16, pp. 41-46.
Masui, A., 1993, "Newly Developed Throwing Application Using the Cycloprothrin U-Granule Package," *Agrochemicals Japan,* No. 63, pp. 2-3.
Noritake, K., 1993, "Labor Saving Application of Jumbo Pellet Type Herbicide," *Agrochemicals Japan,* No. 63, pp. 5-6.
Ohtsubo, T., Fujimoto, I., and Kakuta, Y., 1996, "Baktop MC for Termite Control," *Sumitomo Kagaku,* Vol. 1996-II, pp. 25-32.
Ohtsubo, T., Takeda, T., Tsuda, S., Kagoshima, M., and Tsuji, K., 1989, "Formulation Factors of Fenvalerate Microcapsules Influencing Insecticidal Efficacy and Fish Toxicity," *Journal of Pesticide Science,* Vol. 14, pp. 235-239.
Ohtsubo, T., Takeda, H., Tsuda, S., and Tsuji, K., 1991, "Formulation Factors of Pyrethroid Microcapsules Affecting Rainfastness, Phytotoxicity and Mammalian Toxicity," *Journal of Pesticide Science,* Vol. 16, pp. 413-419.
Ohtsubo, T., Tsuda, S., Kawada, H., Manabe, Y., Kishibuchi, N., Shinjo, G., and Tsuji, K., 1987, "Formulation Factors of the Fenitrothion Microcapsule Influencing the Residual Efficacy against the German Cockroach," *Journal of Pesticide Science,* Vol. 12, pp. 43-47.
Ohtsubo, T., Tsuda, S., and Tsuji, K., 1991, "A Study of the Physical Strength of Fenitrothion Microcapsules," *Polymer,* Vol. 32, pp. 2395-2399.
Ohtsubo, T., and Tsuji, K., 1997, "Controlled Release Formulation," *Pesticide Formulation Guide,* Pesticide Science Society of Japan, Japan Agricultural Formulation and Application Committee, Japan Plant Protection Association, pp. 63-71.
Sekiguchi, M., Takahashi, I., Masui, A., and Kojima, T., 1991, "Formulation Technique of Granules Resurfacing after Submerged Application," *Journal of Pesticide Science,* Vol. 16, pp. 325-334.
Suzuki, I., 1996, "Water-Soluble Package of Pesticides-Pesticide Packaging by PVA Film-," *The 16th Japan Agricultural Formulation and Application Symposium,* pp. 36-48.
Takematsu, T., Morinaka, H., Murakami, M., Miyamura, N., and Nonaka, Y., 1995, *Japanese Patent,* Tokkai Hei 7-47522.
Takeshita, T., 1994, " Labor Saving Granular or Flowable Herbicides for Paddy Rice Field," *Agrochemicals Japan,* No. 64, pp. 4-6.
Tsuda, S., Ohtsubo, t., Kawada, H., Manabe, Y., Kishibuchi, N., Shinjo, G., and Tsuji, K., 1987, " A Way of Action of the Fenitrothion Microcapsule as a Residual Cockroach Control Formulation," *Journal of Pesticide Science,* Vol. 12, pp. 23-27.
Tsuji, K., 1990, "Preparation of Microencapsulated Insecticides and their Release Mechanism," *Controlled Delivery of Crop Protection Agents,* Taylor and Francis, London, pp. 99-122.

Tsuji, K., 1993, " Formulation Technology-Controlled Release Technology in Pesticide Formulation-," *Development of Medicine, Vol. 18, Development of Pesticide I*, Hirokawa Shoten, Tokyo, pp.101-124.

Tsuji, K., 1994, " Agricultural Industry and Surface Active Agents," *Development and Recent Technology of Functional Surface Active Agents*, CMC, pp. 200-224.

Tsuji, K., 1996," Recent Trend of New Pesticide Formulation Development," *Bulletin of Japan Plant Protection Association*, No. 145, April 20, pp. 3-6.

Tsuji, K., 1997a, " Functionalization of Pesticide Formulation," *Powder Science and Engineering*, Vol. 29, No. 4, pp. 33-44.

Tsuji, K., 1997b, "Trend of Formulation Development in Future," *Pesticide Formulation Guide*, Pesticide Science Society of Japan, Japan Agricultural Formulation and Application Committee, Japan Plant Protection Association, pp. 235-236.

Tsuji, K., 1998a, "Recent Trends in Pesticide Formulations," *Formulation Science*, Vol. 1, The Association of Formulation Chemists, pp. 53-83.

Tsuji, K., 1998b, "Recent Trend of Pesticide Formulation Development," *Sumitomo Kagaku*, Vol. 1998-1, pp. 79-90.

Tsuji, K., 1999a, "Application and Particle Design of Insecticide Microcapsules," *Controlled-Release Delivery Systems for Pesticides*, Marcel Dekker, New York, pp. 55-85.

Tsuji, K., 1999b, "Particulate Design of Pesticide Microcapsules," *Atomization*, Vol. 8, No. 23, pp. 140-159.

Tsuji, K., and Ohtsubo, T., 1993, " Granulation in Pesticide Formulations," *Powder Science and Engineering*, Vol. 25, No. 9, pp. 43-49.

Tsuji, K., Shinjo, G., Ito, T., Tsuda, S., and Takahashi, N., 1989, "Sumithion Microcapsule," *Sumitomo Kagaku*, Vol. 1989-I, pp. 4-25.

Ueda, S., and Watanabe, T., 1997, "Jumbo Herbicides," *Pesticide Formulation Guide*, Pesticide Science Society of Japan, Japan Agricultural Formulation and Application Committee, Japan Plant Protection Association, pp. 72-76.

Samir S. Ashrawi,[1] Curtis M. Elsik,[1] Howard M. Stridde,[1] and George A. Smith[1]

Strategies for Agricultural Formulations: A Statistical Design Approach

---

Reference: Ashrawi, S. S., Elsik, C. M., Stridde, H. M., and Smith, G. A., "**Strategies for Agricultural Formulations: A Statistical Design Approach,**" *Pesticide Formulations and Application Systems: A New Century for Agricultural Formulations, Twenty First Volume, ASTM STP 1414,* J. C. Mueninghoff, A. K. Viets, and R. A. Downer, Eds., American Society for Testing and Materials, West Conshohocken, PA, 2001.

Abstract: The traditional approach used by formulators of agricultural and pesticide formulations relies heavily on experience, and has been quite valuable in generating acceptable formulations that may or may not be at a global optimum. However, it can be time consuming, and it generally lacks the ability to provide a global view of the contributions of the different components to the performance of the whole. Additionally, the approach does not permit the use of the collected information to investigate or predict the behavior of formulations within the composition space without carrying out further experimentation.

In the statistical design approach, one begins with the definition of objectives and constraints, then sets up the experiment with the appropriate response factor measurements. After data collection, a best-fit mathematical model is generated to describe the results and can be displayed graphically in various ways for various purposes. Among other things, model analysis typically reveals how each component contributes to the properties being measured. The model is used to locate optimum performance regions that balance the properties of the different components and satisfy any other constraints that are imposed. Scenario analysis can then be used to predict the performance as component levels are changed. In this work, we will discuss the conceptual framework of the statistical design approach, and exemplify its usefulness with an emulsifiable concentrate formulation.

Keywords: experimental design, statistical design, formulation, emulsifiable concentrate, surfactant, surfactant blend, linear alkylbenzene sulfonate, alcohol ethoxylate, alcohol alkoxylate

---

[1]Surfactant Applications Manager, Research Engineer, Project Chemist, and Research Chemist, respectively, Surfactants and Intermediates, The Austin Laboratories, Huntsman Corporation, 7114 North Lamar Blvd., Austin, TX, 78752.

Introduction

Researchers engaged in the development of agricultural and pesticide formulations face a rather laborious and time-consuming task with every new formulation project. They must make a choice among a myriad of raw materials. They must then combine these materials in the proper ratios to obtain the best cost-performance combination, while satisfying a whole host of physical property and stability constraints. Most importantly, they must be sure that the formulation will be robust enough to account for variations of different origins in commercial production, such that the product will meet specifications with minimal or no rework. In reality, formulators are almost always faced with the demand for "something that works," within tight time constraints.

Traditionally, formulators of agricultural and pesticide formulations have relied heavily on experience, instinct and trial-and-error to meet their formulation goals. This approach has been quite valuable in generating acceptable formulations that may or may not be at a global optimum. However, this approach generally lacks the ability to provide a global view of the contributions of the different components to the performance of the whole. Additionally, this approach does not permit the formulator to investigate or predict formulation behavior within the composition space, without carrying out further experimentation. Hence it can be quite time-consuming as one tweaks the formulation to a new composition.

In this paper, we describe a framework for an alternative approach using statistical screening and mixture design techniques to arrive at optimized and robust formulation, and to identify the role of different ingredients in the formulation. Past efforts using experimental design software for agricultural formulations have sometimes ended with conclusions that reported mixed benefits (Butler 1997, Butler and Schultz 1998).

Traditional Formulation Methods

In the traditional trial-and-error approach, the formulator starts out with a set of defined goals for the end product. Using background experience, the formulator arrives at some starting formulation. He then proceeds to vary the individual components in order to obtain acceptable or better than acceptable levels of a particular performance response. The effort continues as a balancing act that is expected to optimize the observed responses, one at a time.

In this Edisonian approach, no bold steps are taken in the component variations, which limits the composition space that is investigated to small perturbations around the starting formulation. Because responses are essentially optimized one at a time, one is never sure that all properties and performance responses have been optimized. Furthermore, one is not sure if the observed optimum is local within the small composition space that was investigated, or global within a much bolder space. Finally, since this approach does not yield maps of performance and properties over the composition space, this approach does not permit any prediction for the behavior without continuous tweaking.

Notwithstanding all their shortcomings, traditional formulation approaches have always yielded workable and valuable solutions that have met the defined goals of performance criteria, physical properties, and cost of finished product.

Statistical Experimental Design

We have used various techniques of statistically designed experiments in our laboratories (Ashrawi and Smith 2000) to help develop and optimize formulations for different applications, including agricultural and pesticide formulations. We find great benefits in using three of those techniques. Factorial Screening Designs are extremely helpful in identifying the vital factors or components that affect the desired product. Response Surface Designs are extremely helpful in locating the ideal criteria or process settings that yield optimum performance. And finally, Mixture Designs yield performance maps over a defined composition space which enable us to discover the optimum formulation and to predict performance in other regions of the defined composition space.

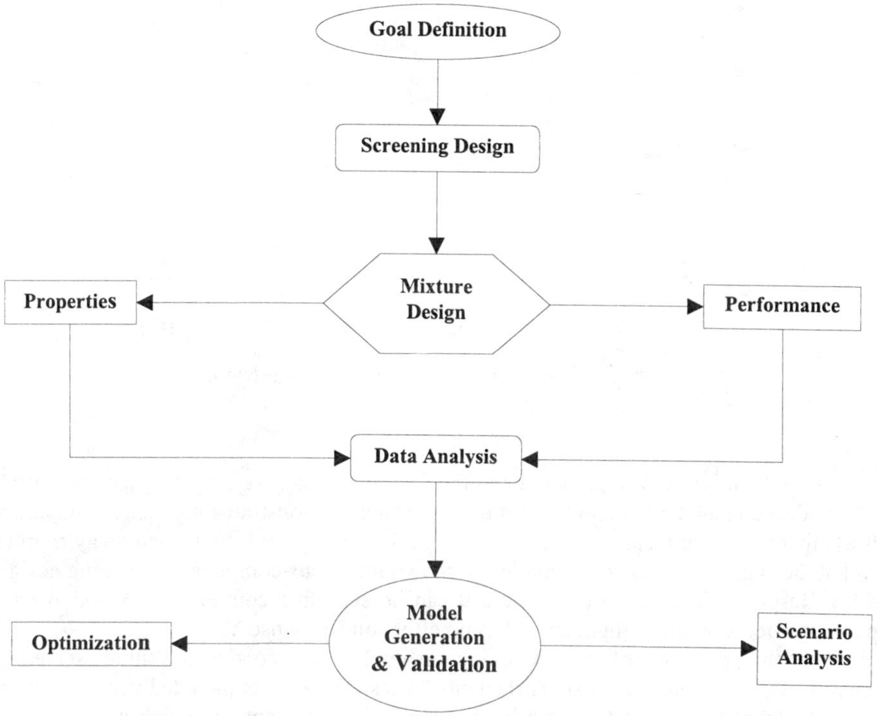

Figure 1 - *Framework for formulation development using statistical experimental design.*

There are several commercially available software packages that enable the formulator to generate designs, analyze data, generate models of best fit and optimize to multiple responses both numerically and graphically.

Figure 1 describes the framework of our statistical experimental approach to formulation development. Beginning with a well-defined goal of achieving certain performance criteria, physical properties and cost, we conduct screening experiments to determine the components that would be vital to achieving our goal.

Figure 2 shows a schematic factorial screening design using two components, A and B. In this design, one prepares at least five mixtures (dark shaded points) that include the apexes of the composition space and a centroid point. The centroid is duplicated to get a measure of the variation in the measurement. Four more points (light shade) can be added to determine any curvature in the response of the performance criteria.

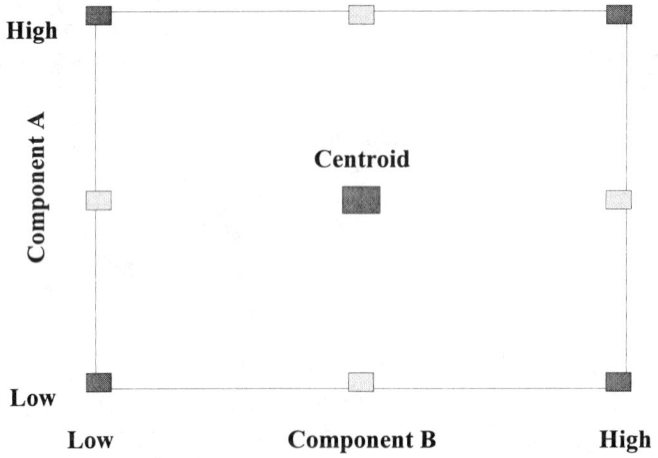

Figure 2 - *Schematic factorial screening design.*

Once the mixtures are prepared and the desired property or performance response is measured, the results can be plotted in a manner that demonstrates how each component affects that particular response as the amount of that component is changed away from its level at the centroid. This is schematically shown for a four-component screening design, with a Response Y, in Figure 3, where it can be seen that components A and B have opposing effects, while component C has no effect on Response Y.

For the purpose of our working example, we conducted three screening experiments, each comprising six surfactants. These surfactants included various salts of linear alkylbenzene sulfonate, and a host of nonionic surfactants in which we used three different hydrophobes, varied the ethylene oxide chain length and included some alkoxylates that contained both ethylene oxide and propylene oxide. All told, we screened 18 surfactants in four families, and selected three components for our next step, the Mixture Design. These surfactants were a linear alkylbenzene sulfonate (LAS), an alcohol alkoxylate (AA) and a linear alcohol ethoxylate (LAE).

Next, we use these three components to develop a mixture design at a fixed total surfactant concentration. The physical properties and performance responses are then measured and recorded.

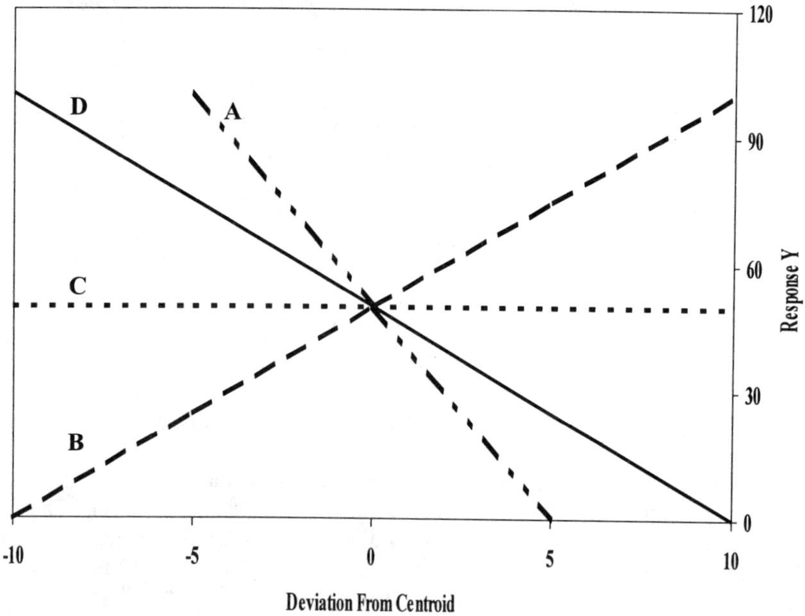

Figure 3 - *Schematic four-component screening results.*

A model that best fits the data is then generated and validated by exploring its statistical significance. With a good-fit model, one can then proceed to optimization of the composition to meet the required performance criteria.

EC Formulation: A Working Example

The goal for this study was to develop a single-blend surfactant system that meets all performance criteria for a 4 lb/gal 2,4-D ethylhexyl ester phenoxy emulsifiable concentrate (EC) formulation with a paraffinic solvent. The surfactants must be 40CFR 180.1001 approved. The performance criteria included a maximum of 2 mL sedimentation after 1 hour at 5 and 25°C with no oil present. The bloom or spontaneity must be 3 or greater on the scale defined in Table 1. This means that no oil must break out of the emulsion cloud that forms spontaneously when the EC is added to water. Hardness was varied from 34 to 342 to 1000 ppm WHO. The dilution rate is 5 mL EC to 95 mL water to study both spontaneity and sedimentation. Design-Expert®[2] was the software used for all experiments.

[2]Design-Expert is a registered trademark of Stat-Ease, Inc., Minneapolis, MN.

Table 1 – *Huntsman spontaneity score for emulsifiable concentrates.*

Spontaneity Score	**Description**
0	Little/No Emulsification
1	Some Emulsification Mostly Free Oil
2	Mostly Emulsion Some (Can be Minor) Free Oil Drops
3	NO Free Oil at All Emulsion Cloud Smooth
4	Rich White Emulsion Heavy Coating on Tube Wall
5	Bluish-White Emulsion Very Heavy Coating on Tube Wall

The surfactant blend was comprised of only three individual commercially available surfactants. The surfactant blend was present in the EC at 4% active level. There were several screening mixture designs run looking at eighteen different surfactant candidates from four different chemistries. The results were used to select the final three surfactant finalists to use in the ensuing principal mixture design.

The final mixture design incorporated a linear alkylbenzene sulfonate (LAS), an alcohol alkoxylate (AA), and a linear alcohol ethoxylate (LAE). As usual, the blend will consist of both anionic and nonionic surfactants to be most stable to a combination of changes in water hardness and temperature. The Design-Expert software defined the ten compositions to evaluate that are listed in Table 2. The LAS was varied between 50 and 75% of the blend. The AA and LAE were studied between 5 and 25%. The candidate surfactant blend compositions are shown on a ternary diagram in Figure 4.

Table 2 – *Experimental design surfactant blend compositions.*

Anionic	**Nonionic**	
LAS	Alcohol Alkoxylate	Linear Alcohol Ethoxylate
70	5	25
75	12.5	12.5
70	20	10
50	25	25
68	16	16
70	10	20
70	25	5
60	20	20
60	25	15
60	15	25

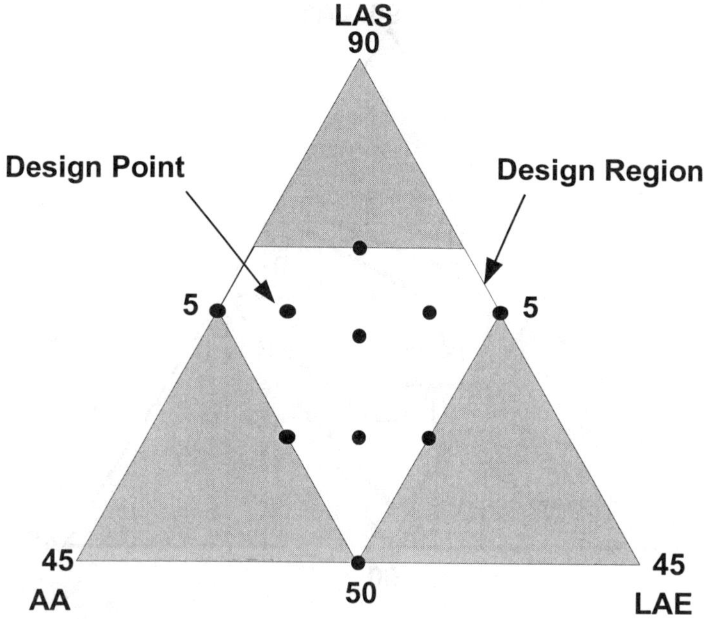

Figure 4 - *Mixture design surfactant blend compositions.*

These surfactant blend compositions were included in the 4 lb/gal 2,4-D EC at an active content of four percent. The experimental design responses were defined as sedimentation and spontaneity at 5 and 25°C, evaluated at water hardness levels of 34, 342, and 1000 ppm WHO. The Design-Expert software can be used to plot responses as traces, contours, or response surfaces. Figure 5 shows the contour lines for spontaneity evaluated at 1000 ppm and 25°C.

Figure 5 shows that spontaneity is maximized at high levels of LAS and LAE with a low level of AA. The desired region where the formulation performs at or above the target spontaneity score of three will be enclosed to the right of the "3.0" contour line.

A similar plot showing the sedimentation response at 1000 ppm and 25°C is illustrated in Figure 6. From this figure it is clear that sedimentation is minimized at high levels of LAE and AA with a low level of LAS. There will be a large region, however, where both spontaneity and sedimentation will meet or beat their target values. Figure 7 combines these two criteria in a two-response optimization as calculated by the experimental design software. The acceptable region is shown as the area that satisfies both of the response targets of spontaneity (> 3) and sedimentation (< 2 mL). Figure 7 shows this area is a significant portion of the original design space when only 1000 ppm water hardness is considered.

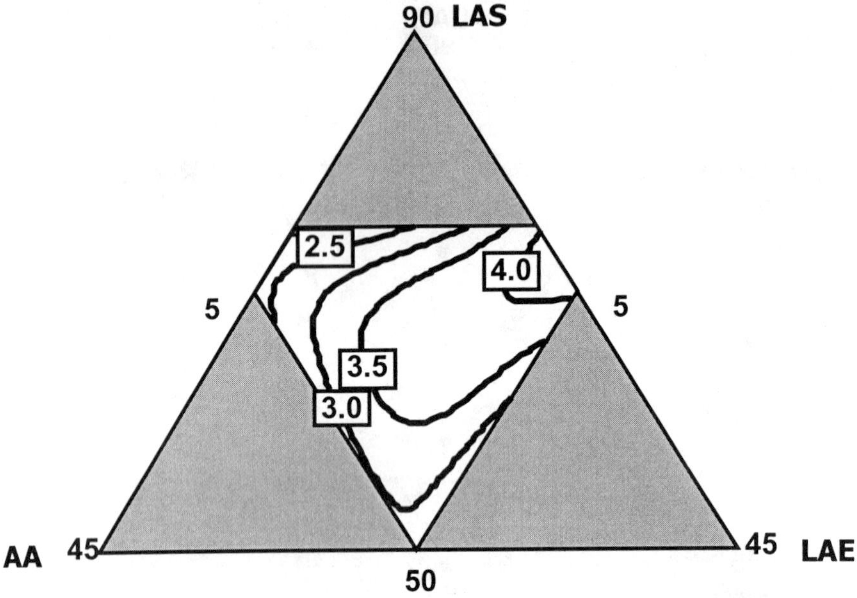

Figure 5 – *Spontaneity response contour lines at 1000 ppm and 25°C.*

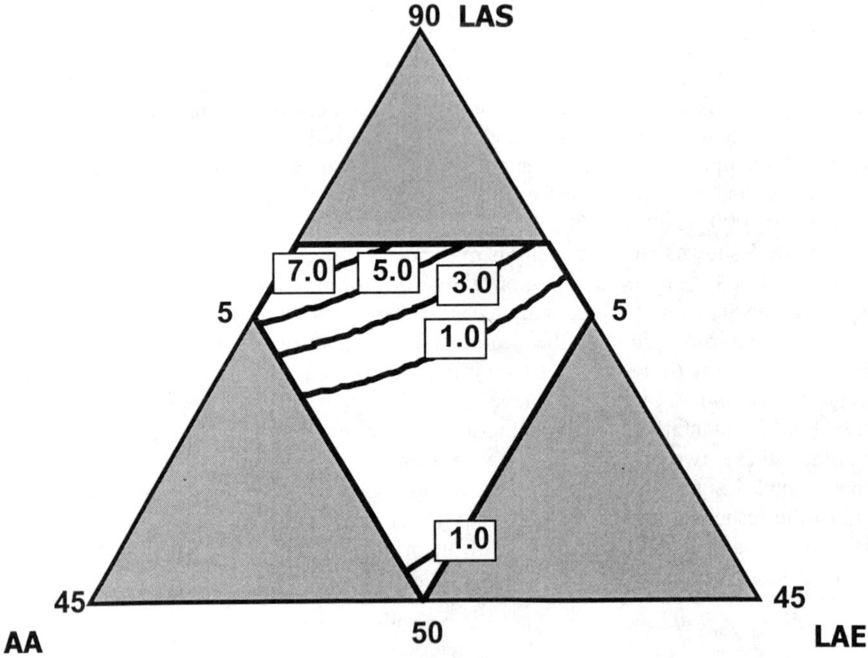

Figure 6 – *Sedimentation response contour lines at 1000 ppm and 25°C.*

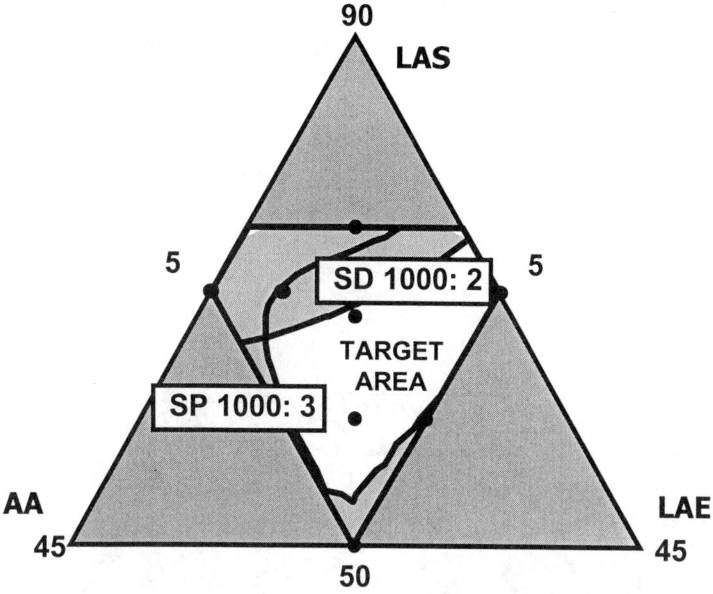

Figure 7 – *Spontaneity and sedimentation optimization at 1000 ppm and 25°C.*

The design software can combine all three water hardness levels studied into one contour plot that contains both spontaneity and sedimentation. Figure 8 shows five of these six contours plotted for 25°C. Sedimentation was less than 2 mL at 34 ppm water hardness for the entire design space, so the contour for this response does not show up in Figure 8. The resulting acceptable surfactant blend composition area is roughly defined by a triangle with high levels of AA and intermediate levels of LAS and LAE. The fact that the original design space completely encompasses this optimum area is significant. If the optimum area was not bounded by the original design, then the best reaction would be to repeat the mixture optimization design with the new surfactant compositions to make sure the model's mathematical predictions were robust.

The design software also has a function that allows the formulator to determine the actual composition at different locations in the ternary diagram. A flag planted in the middle of the optimum region defined in Figure 8 yields a surfactant blend composition of 63% LAS, 21% AA, and 16% LAE. This composition should yield the most robust EC formulation possible, without considering economics into the equation. Note that for both LAS and LAE the optimum level is almost the exact average of the extreme ranges defined for the mixture design (62.5 and 15%). This shows the value of having an experienced formulator involved in setting up the original experimental design.

In the real world the optimum found in Figure 8 for 25°C would be combined with an optimum found for 5°C to yield the final surfactant blend composition. This combined optimum should provide the most robust EC formulation possible from a technical performance perspective. Further optimization of the surfactant composition could also be performed using the final cost of the blend as a response if so desired.

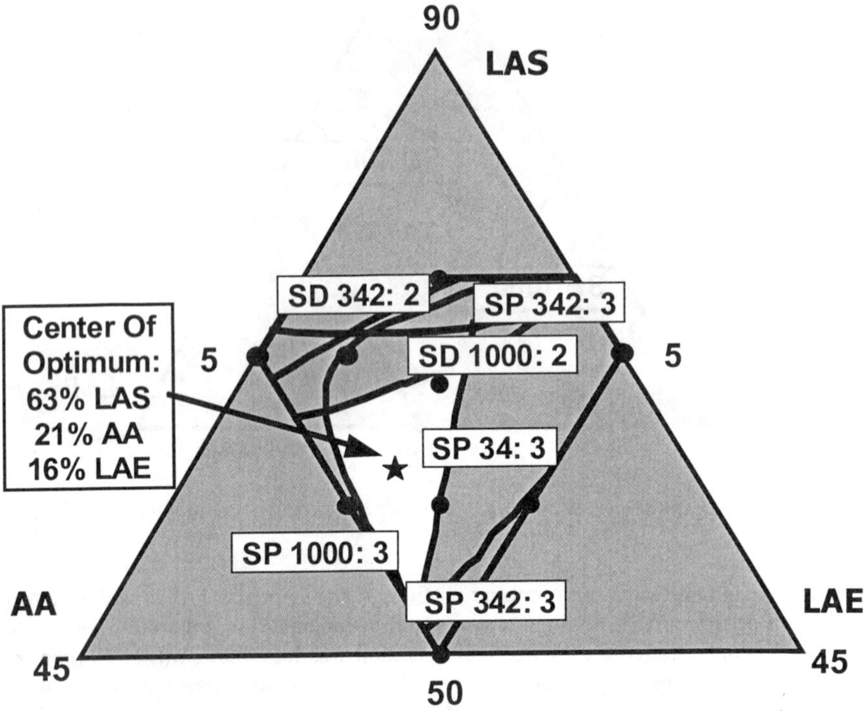

Figure 8 – *Spontaneity and sedimentation optimization for 34, 342, and 1000 ppm water.*

Conclusions

The methodology that we employed in this work is very valuable because it minimizes experimentation, elucidates component effects, yields an optimum operating region with a global optimum and a clear understanding of cost – performance relationships. The methodology requires good goal definition as well as meaningful and quantifiable response measurements.

In our working example, we used this methodology with a Mixture Design to develop a 3-component surfactant blend for a phenoxy EC formulation. The optimum formulation was robust to water hardness, temperature and technical lot-to-lot variation.

This strategy can also be used to develop products where processing conditions have an influence on final performance such as Suspension Concentrates, Concentrated Emulsions, Suspoemulsions, and Water Dispersible Granules. In these cases instead of using a Mixture Design, a Response Surface Design would be used with the particular processing variables used as the input factors.

Acknowledgments

The authors wish to acknowledge the Huntsman Corporation for permission to publish this work.

References

Ashrawi, S. S., and Smith, G. A., 2000, "Use of Experimental Mixture Design to Optimize Cleaning Formulations," *Proceedings of the 5th World Surfactants Congress*, May 29 – June 2, 2000, pp. 1253-1261.

Butler, B. J., 1997, "Development of Emulsifiable Concentrate Formulations Using Experimental Design Software," *Pesticide Formulations and Application Systems: Seventeenth Volume, ASTM STP 1328*, G. R. Goss, M. J. Hopkinson, and H. M. Collins, Eds., American Society for Testing and Materials, West Conshohocken, PA, pp. 11-22.

Butler, B. J., and Schultz, A. K., 1998, "Developing Emulsifiable Concentrate Formulations Using Droplet Size and Experimental Design Software," *Pesticide Formulations and Application Systems: Eighteenth Volume, ASTM STP 1347*, J. D. Nalewaja, G. R. Goss, R. S. Tann, Eds., American Society for Testing and Materials, West Conshohocken, PA, pp. 168-184.

Liane Stocky,[1] Dan Cantalupi,[2] John Lucas,[3] and Ralph Kowalik[4]

Formulating Options for Aerosol Insecticides

Reference: Stocky, L., Cantalupi, D., Lucas, J., and Kowalik, R., **"Formulating Options for Aerosol Insecticides,"** *Pesticide Formulations and Application Systems: A New Century for Agricultural Formulations, Twenty First Volume, ASTM STP 1414,* J. C. Mueninghoff, A. K. Viets, and R. A. Downer, Eds., American Society for Testing and Materials, West Conshohocken, PA, 2001.

Abstract: With ever tightening VOC regulations for consumer insecticides, formulators are being pushed to use more water and/or more LVP (low vapor pressure) fluids that are not reportable as VOCs in their products. The particular choice of LVP fluid and its concentration may also affect the biological performance of insecticide formulations. To demonstrate these effects, a variety of hydrocarbon LVP carrier fluids were combined with the insecticides S-bioallethrin[5] and deltamethrin[6] in both solvent and water-based insecticide formulations. Experiments were conducted to evaluate knockdown and residual activity against German cockroaches.

Data indicated that use of hydrocarbon LVP carrier fluids with higher levels of normal paraffins resulted in faster knockdown of the cockroaches under a variety of test conditions. Higher concentrations of hydrocarbon LVP carrier fluids in water-based formulations also enhanced their knockdown performance. The residual activities of water-based formulations, inferred from the lengths of time the formulations remained efficacious after they were applied to ceramic tile and painted plywood, were adversely affected by higher concentrations of normal paraffins or LVP hydrocarbons.

The results suggest that normal paraffins and higher concentrations of LVP hydrocarbon fluids may increase the penetration of the formulations into insect and household surfaces. With the insects this leads to faster intake of insecticide. With the household surfaces this reduces insecticide availability. One can use this information to select active ingredients, LVP fluids, and their concentrations, along with other formulants, to optimize formulation performance for specific application requirements and meet VOC regulations. Of the LVP hydrocarbon fluids tested, a mixed aliphatic fluid with a moderate concentration (20% w/w) of normal paraffins appeared to offer the best balance of properties for use in crawling insect killers.

---

[1]Associate Biologist, [2]Formulation Chemist, and [3]Biology Manager and GoldCrest® Products Advisor, all at Aventis Environmental Science, 95 Chestnut Ridge Road, Montvale, NJ 07645.

[4]Product Application Specialist, ExxonMobil Chemical Company, 5200 Bayway Drive, Baytown, TX 77520.

[5]Esbiol® and [6]DeltaGard®. GoldCrest®, Esbiol®, and DeltaGard® are registered trademarks of the Aventis Group, Montvale, NJ.

Keywords: aerosols, cockroaches, deltamethrin, insecticides, knockdown, LVP hydrocarbon fluids, S-bioallethrin

Background

Over the past decade consumer aerosol insecticides have been subject to VOC (volatile organic compound) regulations that limit the amounts and/or types of solvents and propellants used in formulations. The regulations started in California and spread to a few other states. In 1998 the US Environmental Protection Agency enacted a national rule, and more restrictive VOC limits are still coming in California and a group of Northeastern states. To meet the regulatory challenges and maintain product efficacy, formulators have increasingly relied on LVP (low vapor pressure) hydrocarbon carrier fluids that have vapor pressures less than 13.3 Pa (0.1 mm Hg) at 20 °C or initial boiling points greater than 216 °C. These fluids are not reportable as VOCs in regulated consumer products due to their low volatility and reduced atmospheric availability, and they have been used in both solvent and water-based systems. We have found that the LVP fluid's specific composition, and its concentration in water-based systems, can influence the knockdown of insects and the residual activity of the insecticide after it is sprayed onto household surfaces. This paper demonstrates the effects in crawling insect killer (CIK) formulations against German cockroaches and suggests formulating ideas based on them.

Literature Perspective

The effects of solvent concentration and hydrocarbon fluid composition on insect knockdown have been reported to a limited extent. A study of water-based aerosol insecticide formulations suggested that 20% w/w concentrations of an isoparaffinic hydrocarbon fluid provided the fastest knockdown of houseflies and mosquitoes (Yamaguchi et al. 1976). The highest hydrocarbon concentration tested was 20%, and 10% concentrations provided nearly equivalent knockdown performance as 20%. Lower (0% and 5%) hydrocarbon concentrations were inferior.

A later study of solvent-based aerosol insecticide formulations showed that kerosene formulations with higher concentrations of normal paraffins provided faster knockdown of houseflies, mosquitoes, and cockroaches (Tsuda and Okuno1985). The same study suggested that among single isomer normal paraffins, those with a chain length of about 14 carbon atoms provided the fastest knockdown of cockroaches.

A third study of solvent- and water-based aerosol insecticides concluded that solvent volatility did not affect the knockdown of houseflies if the data were normalized to the dose of active ingredient that reached the fly (White et al. 1992). They found that solvent volatility or boiling range could affect the insecticide spray characteristics and the contact of insects and aerosol droplets, but insect knockdown depended solely on the amount of active ingredient delivered to the insect and not on the solvent that carried it.

There are apparent gaps in the literature regarding the effects of hydrocarbon type and concentration on the knockdown of crawling insects by water-based insecticides. One paper suggests that for water-based aerosols the variety of kerosene has an effect on efficacy to a certain degree, but this effect is not great because of the small amount of kerosene commonly used (Tsuda 1986).

Experiments

Materials and Test Methods

The experimental methods compared the knockdown and residual activity of solvent- and water-based aerosol insecticides. The insecticides contained four different LVP hydrocarbons that varied in chemical composition. Properties of the four LVP hydrocarbon fluids are listed in Table 1. The four fluids have similar boiling ranges but have notable differences in specific paraffin contents.

Table 1 – *LVP Hydrocarbon Fluid Properties*

Fluid	Paraffin, % w/w			Boiling Range,°C
	Normal	Iso	Cyclo	
A	100	0	0	222-243
B	20	45	35	224-238
C	1	32	67	227-288
D	0	75	25	223-254

The pyrethroids S-bioallethrin and deltamethrin were used as the knockdown and kill agents, respectively. They were used at concentrations of 0.01 or 0.05% w/w S-bioallethrin and 0.01% deltamethrin in both solvent- and water-based formulations. The 0.05% w/w S-bioallethrin and 0.01% w/w deltamethrin levels are typical of those used in commercial aerosol insecticides. The 0.01% w/w S-bioallethrin level allows hydrocarbon fluid effects on knockdown to be more readily discriminated. Both active ingredients were fully soluble in the four LVP hydrocarbon fluids at the concentrations tested.

Solvent-based formulations contained the active ingredients, the LVP fluid, and propellant. Water-based formulations contained either 10, 20, or 30% w/w of one of the LVP hydrocarbon fluids, about 1% w/w surfactants and additives, water, and propellant. All formulations were packed in standard 237 ml (8 oz) aerosol cans equipped with standard CIK valves and charged with 19% w/w A-46 propellant (a mixture of 81 mol% isobutane and 19 mol% propane that produces a nominal aerosol pressure of 317 kPa (46 psi) at 21 °C (70 °F)).

Two- to four-week-old, male, German cockroaches (*Blattella germanica*) were the target insects. Their knockdown and mortality responses to the insecticides were measured via 1) direct spray 2) wet walk over and 3) residual testing techniques.

Direct Spray - In the direct spray technique a target dose of 1 gram of formula was sprayed on 10 insects from a distance of 0.46 m (18 in). The insects were then transferred to clean jars and observed for knockdown. Cockroaches were considered knocked down when they were on their backs with their legs in the air and unable to right themselves. Very occasionally insects became incapacitated in an upright position. If this was observed, forceps were immediately used in an attempt to flip these insects onto their backs to determine if they were knocked down or not. Four replicate sprays were conducted for each formulation tested. The water-based systems with 20 and 30% LVP hydrocarbon fluid produced an unsatisfactory pin stream spray pattern when they were formulated and packaged according to the basic recipe for 10% LVP fluid water-based systems. These formulations were not tested via direct spray but rather via the wet walk over technique.

Wet Walk Over - In the wet walk over procedure about 1 gram of formulation was sprayed onto individual ceramic tiles. 10 cockroaches were then placed on top of each wet tile and kept there for 1 minute. The roaches were then transferred to clean jars and observed for knockdown. Five replicates were conducted for each formulation tested. This technique was used for all water-based formulas. KT50 and KT90 statistics, the times for 50 and 90% of the roaches to be knocked down, were calculated from the data from both direct spray and wet walk over methods using probit analysis. Results averaged over the replicates were also plotted as % knockdown vs. time.

Residual - Painted plywood and unglazed ceramic tile surfaces were used to compare the residual activity of several water-based formulations. The plywood was painted with 2 coats of white semi-gloss latex paint and allowed to dry for at least 2 weeks before it was treated with insecticide formulations. One gram of formulation was sprayed onto each substrate and allowed to dry for 24 hours. Residual testing was then conducted at 24 hours and 1 week after treatment on both substrates and also at 2 and 3 weeks after treatment on the ceramic tile. There were five replicates for each formulation tested. For each test, 10 male roaches were placed on the treated surface and confined there using a plexi-glass arena for either 1 minute or 30 minutes. After the timed exposure, the roaches were then placed in clean jars and observed for knockdown over the first 60 minutes and then for mortality at 24 hours. Higher percentage of mortality and longer duration of effectiveness indicate superior residual activity.

Results and Discussion

Knockdown data is displayed in Figures 1 through 4. Summary data is listed in Table 2. Regressed knockdown times KT50 and/or KT90 values are compared for various formulations and application techniques in Figures 1 and 2 and Table 2. Figures 3 and 4 show actual knockdown results averaged over the replicate runs. The data suggests that knockdown is generally faster with 1) higher levels of S-bioallethrin, 2) higher levels of LVP hydrocarbon fluid in the formulation, and 3) higher levels of normal paraffins in the LVP hydrocarbon fluid. Faster knockdown also occurred in the wet walk over tests relative to the direct spray tests on the same formulation. Cockroach mortality 24 hours after contact with insecticide was 100% for all direct spray and wet walk over tests. Uncertainties in the knockdown times, expressed as 95% confidence interval total widths, were approximately $0.7 + 0.35 \cdot KT50$ and $1.4 \cdot KT90$ (minutes).

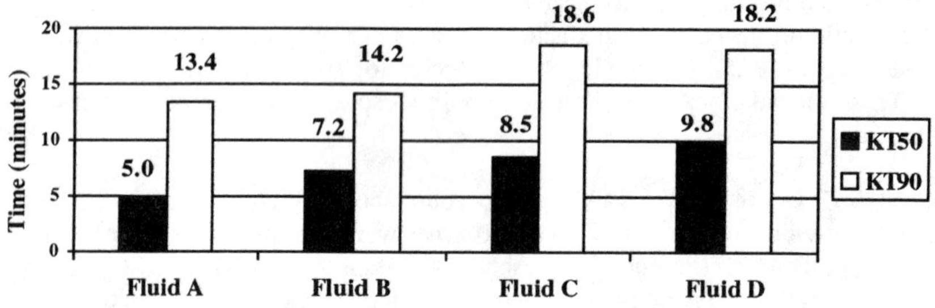

Figure 1 – *Statistical Knockdown of German Cockroaches by Direct Spray of Water-Based Formulations Containing 10% w/w LVP Hydrocarbon Fluid, 0.01% w/w S-bioallethrin, and 0.01% w/w Deltamethrin*

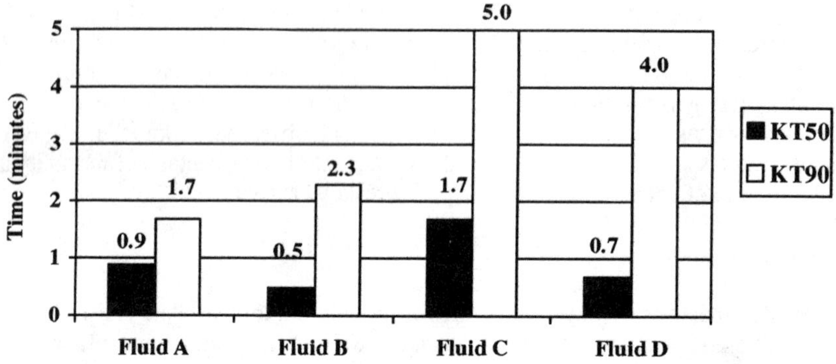

Figure 2 – *Statistical Knockdown of German Cockroaches by Wet Walk Over of Water-Based Formulations Containing 10% w/w LVP Hydrocarbon Fluid, 0.01% w/w S-bioallethrin, and 0.01% w/w Deltamethrin*

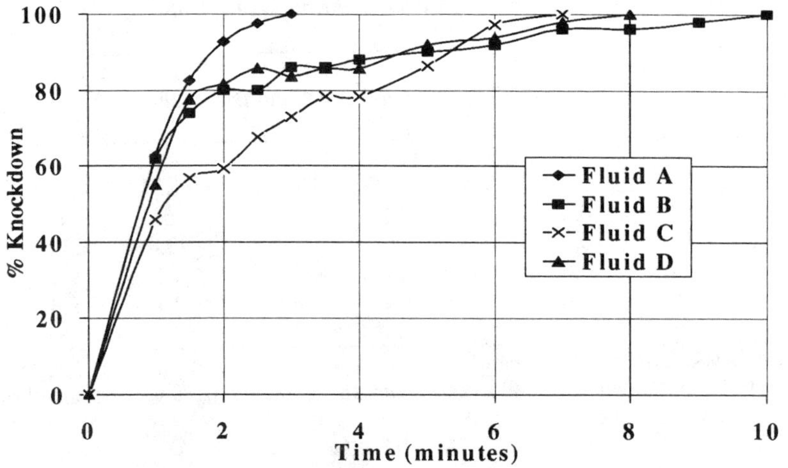

Figure 3 – *Plotted Knockdown of German Cockroaches by Wet Walk Over of Water-Based Formulations Containing 10% w/w LVP Hydrocarbon Fluid, 0.01% w/w S-bioallethrin, and 0.01% w/w Deltamethrin (from same data used for Figure 2)*

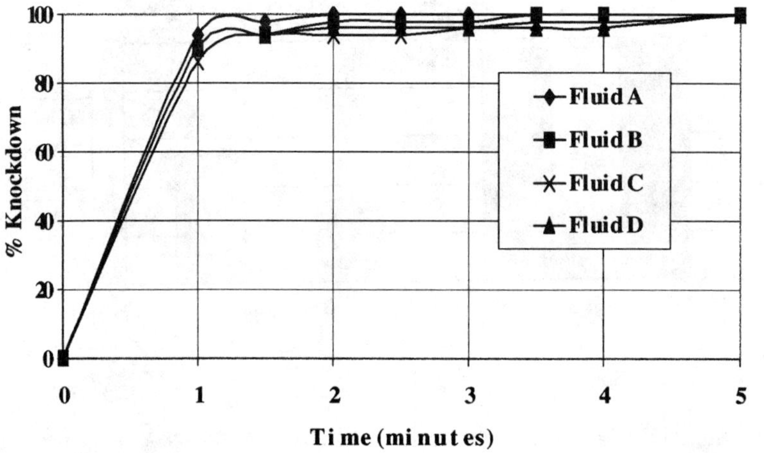

Figure 4 – *Plotted Knockdown of German Cockroaches by Wet Walk Over of Water-Based Formulations Containing 30% w/w LVP Hydrocarbon Fluid, 0.01% w/w S-bioallethrin, and 0.01% w/w Deltamethrin*

Table 2 – *Summary of Cockroach Knockdown Data*

Formula Type[7]	HC Fluid % w/w	Active[8] % w/w	Test Type[9]	Knockdown, KT90 (minutes) Fluid A	Fluid B	Fluid C	Fluid D
WB	30	0.01	WW	0.9	1.2	1.3	1
WB	20	0.01	WW	1.3	1.2	1.5	1.5
WB	10	0.01	WW	1.7	2.3	5	4
WB	10	0.01	DS	14.4	16.3	22.4	25.5
SB	100	0.01	DS	2.3	4.9	4.7	9.7
SB	100	0.05	DS	1.6	1.6	3.1	2.4

[7]WB = water-based, SB = solvent-based
[8]S-bioallethrin knockdown agent (all formulations also contain 0.01% w/w deltamethrin)
[9]WW = wet walk over, DS = direct spray

Figures 5 to 7 show results of the residual test for several water-based formulations. Each of the three graphs shows the percentage of cockroaches that were dead 24 hours after they were exposed to a treated surface. Variables include substrate type, time between treatment and exposure, duration of exposure, and type and concentration of LVP fluid used in the formulation. The results suggest that higher levels of LVP fluids, higher levels of normal paraffins in the LVP fluid, and the more porous substrate (painted plywood) all reduced the time that the treated surfaces were effective against cockroaches.

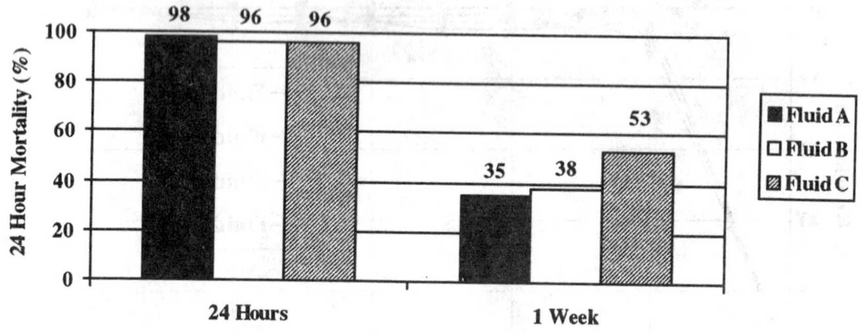

Figure 5 – *Residual Effectiveness of a Water-Based Formulation Containing 10% LVP Hydrocarbon Fluid, 0.01% w/w S-bioallethrin, and, 0.01% w/w Deltamethrin Following 30 Minute Exposures on Painted Plywood, 24 Hours and 1 Week After Treatment*

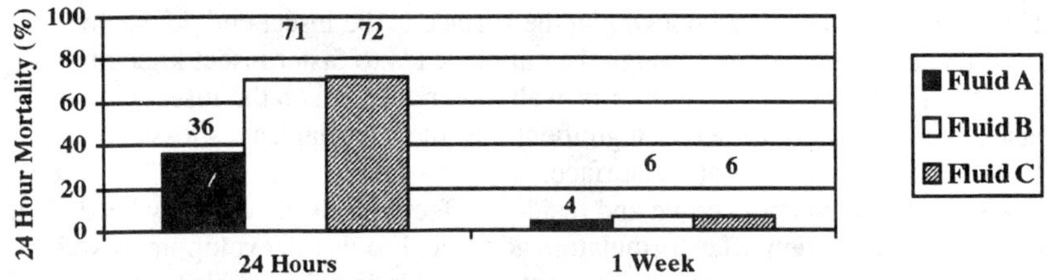

Figure 6 – *Residual Effectiveness of Water-Based Formulation Containing 30% LVP Hydrocarbon Fluid, 0.01% w/w S-bioallethrin, and 0.01% w/w Deltamethrin Following 30 Minute Exposures on Painted Plywood, 24 Hours and 1 Week After Treatment*

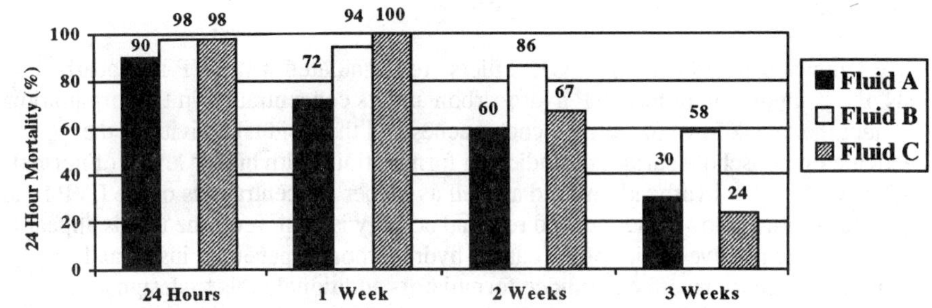

Figure 7 – *Residual Effectiveness of a Water-Based Formulations Containing 20% LVP Hydrocarbon Fluid, 0.01% w/w S-bioallethrin, and 0.01% w/w Deltamethrin Following 1 Minute Exposures on Ceramic Tile, 24 Hours to 3 Weeks After Treatment*

Summary

The present crawling insect killer data demonstrates that higher levels of normal paraffins in solvent-based aerosol insecticides can provide faster knockdown. The trend also applies to water-based emulsion systems containing 10 to 30% w/w LVP hydrocarbon fluids. This data further indicates that concentrations of LVP fluids higher than the typical 10% w/w used in many commercial water-based formulations can lead to faster knockdown. Use of higher concentrations of LVP hydrocarbon fluids does not increase the reportable VOC content of consumer insecticide formulations nor does it change their Class 1 aerosol flammability rating if the LVP fluid concentrations are kept below about 30% w/w. The higher concentration levels, however, may require modification of the typical emulsifier/additive packages and/or valve hardware used to provide appropriate spray characteristics.

The higher levels of normal paraffins or LVP fluid concentrations that enhance knockdown also decrease the residual effectiveness of crawling insect killers. The mechanism for both appears to be faster penetration of insecticide formulation ingredients through a surface. For knockdown the surface is the insect cuticle. Faster penetration of the active ingredient through the cuticle leads to faster insect knockdown. For residual activity the surface is the floor or wall material to which the insecticide is applied. Faster penetration of the active ingredient into the material leaves less active ingredient available to kill insects at the surface.

The tradeoffs between knockdown and residual effectiveness provided by LVP fluid composition and concentration offer formulators additional tools to develop improved products. The LVP fluid composition and concentration can be adjusted along with the other ingredients to accentuate a particular effect or to deliver appropriately balanced performance. Mixed aliphatic LVP fluids with about 20% w/w normal paraffins appear to offer an excellent balance of fluid effects with knockdown close to that of normal paraffins and residual effectiveness comparable to that of isoparaffin/cycloparaffin mixtures.

Conclusions

When water-based crawling insect killers are formulated with LVP hydrocarbon fluids, the composition of the LVP hydrocarbon and its concentration in the formulation can affect the knockdown of German cockroaches and the residual activity of the insecticide on household surfaces. Studies on formulations with higher levels of normal paraffins in the LVP hydrocarbon fluid as well as higher concentrations of the LVP fluid suggest faster knockdown and reduced residual activity is achieved. The trends appear to be related to the relative ability of the carrier hydrocarbon to penetrate insect and household surfaces. The effects provide formulators additional tools to design insecticides with particular attributes. When formulated with the appropriate insecticides, mixed aliphatic carrier fluids with about 20% normal paraffins provide an excellent balance of knockdown activity and residual performance.

References

Tsuda, S., 1986, "Formulation Factors Influencing the Efficacy of Aerosol Insecticides," *Sumitomo Pyrethroid World,* No. 6, pp. 2-4.

Tsuda, S. and Okuno, Y., 1985, "Solvents and Insecticidal Efficacy of the Aerosol Containing Tetramethrin and d-Phenothrin," *Journal of Pesticide Science*, Vol. 10, pp. 621-628.

White, A. W. C., Martin, R., Stewart, D. C., and Wickham, J. C., 1992, "Assessing the Performance Against Houseflies of Indoor Aerosol Space Sprays. Part I. Factors Examined During Tests with Houseflies in Free Flight," *Pesticide Science*, Vol. 34, pp. 153-162.

Yamaguchi, T., Okuno, Y., and Fujita, Y., 1976, "Investigation of Water Based Insecticidal Aerosols," *Aerosol Report*, Vol. 15, No. 9, pp. 255-264.

Ingomar W. Doering[1] and Humbert T. Delli Colli

The Physical Chemistry of Water Dispersible Granules, Part II: The Influence of Dispersant Molecular Weight and Paste Rheology on the Physical Properties of Extruded WDGs

---

Reference: Doering, I. W., and Delli Colli, H. T., **"The Physical Chemistry of Water Dispersible Granules, Part II: The Influence of Dispersant Molecular Weight and Paste Rheology on the Physical Properties of Extruded WDGs,"** *Pesticide Formulations and Application Systems: A New Century for Agricultural Formulations, Twenty First Volume, ASTM STP 1414,* J. C. Mueninghoff, A. K. Viets, and R. A. Downer, Eds., American Society for Testing and Materials, West Conshohocken, PA, 2001.

Abstract: The molecular weight distribution and the degree of sulfonation of Kraft lignin dispersants are used to determine performance plateau regions for the most favorable physical properties of WDG formulations produced by low pressure extrusion. Special emphasis is placed on the preparation of extruder feedstock pastes and on the related rheological properties of such pastes. Their ideal water requirement was determined by torque rheometry and other methods such as yield by deformation. A model study containing blends of high and low degree of sulfonation Kraft lignin dispersants with correspondingly high and low molecular weight fractions demonstrates the existence of such performance plateau regions which permit the selection of a final formulation with a broad spectrum of desirable physical properties. Porosity data are utilized to further demonstrate performance differences within the described platform regions.

Keywords: formulation development, sulfonated Kraft lignin, degree of sulfonation, molecular weight distribution, extrusion, paste rheology, yield value, die expansion, pore size distribution, shelf life, dispersion, dispersional stability

Extruded water dispersible granules are manufactured by a multi-step process that depends upon aspects of a complex physicochemical system to produce a hard crush and attrition resistant particle which is sufficiently large not to pose a dust inhalation hazard and at the same time readily and completely disperses to form a thermodynamically stable dispersion.

The process can be described as follows in Table 1.

---

[1]Industry specialist and manager, respectively, Agricultural Chemicals, Westvaco Corporation, Polychemicals Department, P.O. Box 70848, Charleston, SC, 29415-0848.

Table 1 - *The extrusion process revisited.*

A) Premix with narrow particle size distribution →
B) Discontinuous distribution of water →
C) High paste flow →
D) Maximum dye plate expansion →
E) Low surfactant viscosity @ $\gamma = 10^{-3} - 10^{-1}\ s^{-1}$ →
F) Pores filled with surfactant or large lenses of concentrated surfactant solution →
G) Drying rate selected to minimize capillary pressure and surface tension effects,
H) Finished product with:

High density, high hardness, rapid and complete disintegration, stable dispersion

In WDG technologies as in other modern formulation systems we observe a series of multiple interactions at various levels such as: raw material influences, surfactant influences, process variables and aging phenomena. Some of our earlier work focused on the optimization of dispersibility, hardness and dispersional stability with respect to dispersant molecular weight (Delli Colli and Doering 1999).

The present work introduces torque rheometry and yield measurements by deformation as tools to control preparation and performance of extruder feed stock pastes as well as related physical properties of the final product after drying.

While Leuenberger measured the power consumption profile of mixers during the preparation of wetted powder masses, identifying different degrees of saturation (Leuenberger and Bier 1990) torque rheometers measure the torque signal originating from the mixing device (Parker and York 1990, Parker and Rowe 1990, Chatlapalli and Bhagwan 1998).

Three different signals are usually recorded during a defined period of mixing time: mean torque, mean amplitude and maximum height.

The mean torque is the average torque exerted on the mixer bowl by the wet powder mass at any measurement (log-in) period during mixing.

The torque amplitude is the average of the range of oscillations of torque exerted by the wet mass on the mixer bowl during any measurement period during mixing.

The maximum torque height is the mean of maximum torque observed during any log-in period.

Some of the information that can be isolated by this method:

a) Binder / substrate interactions,
b) Degree of liquid spreading and wetting,
c) Extruder mass consistency,
d) Ideal blending time and
e) Critical liquid requirement of a powder mass for extrusion.

While for most of the above criteria the determination of both mean torque and mean amplitude is sufficient, moistened powder masses for extrusion require that we monitor the maximum height torque as well.

This signal is valuable in predicting the ease of extrusion of moistened powder masses as well as expected levels of power consumption of the extruder and die pressure within the extruder dome.

Based upon our experience and comparison between laboratory and industrial scale, mean torque and mean amplitude signals should not exceed 2 Nm and maximum height peaks should not be significantly greater than 3-3.5 Nm. At levels greater than 4 Nm extruder pastes are usually very hard, not deformable or even dilatant.

If mean torque and mean amplitude values significantly exceed values of 2.5 or 3 Nm dispersion problems are likely due to a stronger binder/substrate interaction. Generally less elastic recovery (die plate expansion) is present in those systems as well.

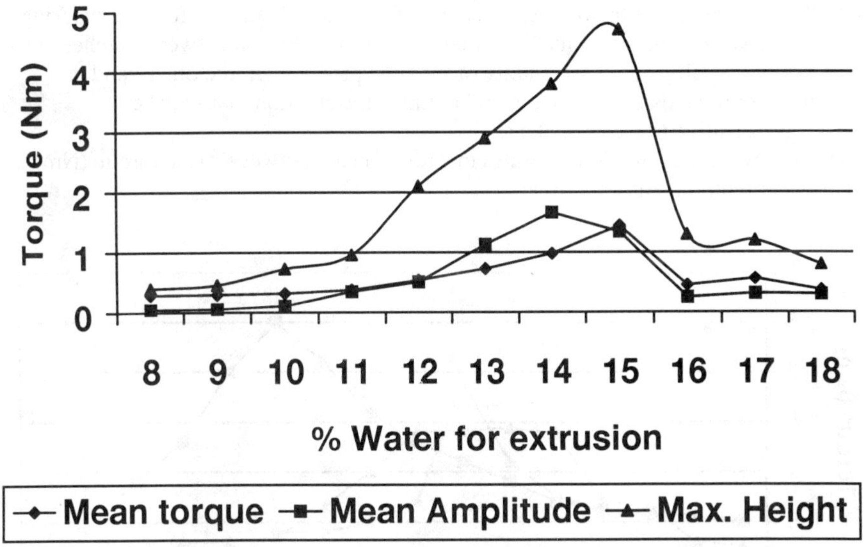

Figure 1 - *Importance of the maximum height peak.*

Although both mean torque and mean amplitude did exhibit acceptable values, the formulation presented in Figure 1 produced a very hard paste characterized by high power consumption and low throughput rates and can be classified as: very difficult to extrude as predicted by the maximum peak height curve.

One of the capabilities of the torque rheometers is the ability to determine the "critical liquid requirement of a moistened powder mass for extrusion".

By definition, the critical liquid requirement of a powder mass, representing the ideal amount of water for low pressure extrusion is a function of the physical properties of the substrate, such as particle size, size distribution, surface area, bulk and tapped density and interfacial properties such as surface kinetics, wetting, and spreading phenomenon.

In this measurement mode the instrument is programmable to receive consecutive water additions to the powder mass filled into a mixing device. As previously described by Leuenberger, energy consumption of a mixer/kneader motor is proportional to shear for a given frequency of the impeller blades. (Leuenberger and Bier 1990)

With increasing water addition more and more void space in the powder mass is being filled with liquid. As a consequence of different degrees of saturation of the powder feedstock a plateau region can often be observed where no further increase of the torque signal is detected.

A broad plateau region in formulations for extrusion is not only helpful in controlling hardness or attrition resistance of WDGs but offers a comfortable working region for process engineers. Small differences of water addition between batches would not appreciably affect the quality of the final product. On the other hand, a very narrow or nonexistent plateau would define a formulation that may be excessively sensitive to water content.

The following diagram demonstrates this relationship between mean torque (Nm) and % water requirement for extrusion.

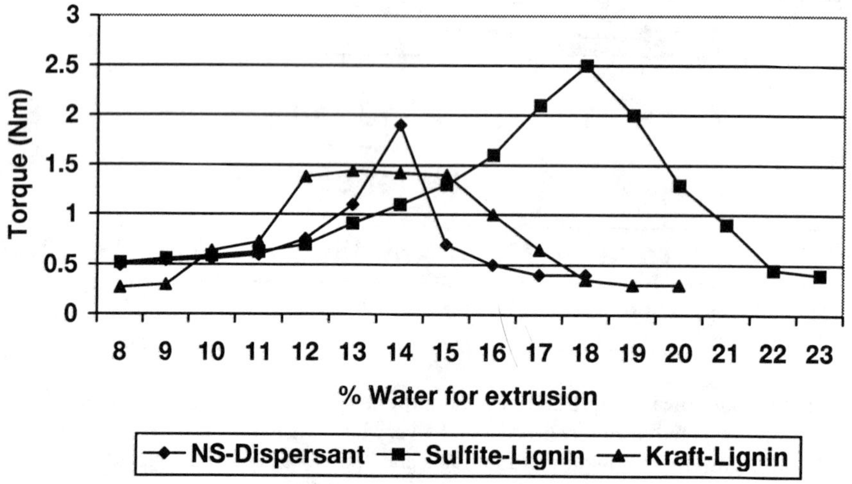

Figure 2 - *Plateau Region – Chlorothalonil 80 WDG formulated with three different dispersants.*

The dispersants of the naphthalene sulfonate and sulfite lignin families showed very narrow peaks at different water levels while the sulfonated Kraft lignin dispersant presented a broad plateau.

The formation of such plateaus in general depends strongly upon the rheological matrix formed at this stage of semi saturated powder masses. The degree of saturation varies and extends over a range of 60-75% for most of our formulations and for most of the literature examples (Leuenberger and Bier 1990).

A contributing factor for the extent of the mean torque plateau region could be the swelling capacity of the formulation when the plateau level is reached. Additional water would not affect deformation related properties, as long as we do not go beyond the "capillary" stage of saturation.

This could as well be the reason why the yield method by deformation works in some examples better than in others.

For formulations exhibiting appreciable levels of elastic deformation and recovery (G'/G"= >1:4) it can be assumed that the mean plateau torque range becomes broader as G' increases.

G'= elastic or storage modulus
G"= viscous or loss modulus

Model Experiment

In order to investigate further the influences of dispersant blends based upon high and low degree of sulfonation Kraft lignin, a series of experiments was conducted using the following experimental design:

A) Active Ingredient: Sulfoanilide analog
B) Dispersant blend: D1 = low degree of sulfonation Kraft lignin
D2 = high degree of sulfonation Kraft lignin
C) Total dispersant concentration: 12%
D) Blends: D1: D2 = 0 : 12 down to D1 : D2 = 12 : 0 in 2 % intervals.
E) Wetting agent: 2% (Chemical class: Naphthalene Sulfonate, Na salt)
F) Mineral clay: qs ad 100%)
G) Active Ingredient concentration: 75%
H) Extruder: LCI Lab bench top basket low pressure extruder (fill volume: 100 ml)
I) Dryer: Buechi laboratory fluid bed dryer
J) Drying temperature: 45- 50°C
K) Residual moisture content: 1 – 2%

Table 2 - *Influence of dispersant molecular weight and degree of sulfonation on suspension, dispersion and attrition resistance.*

Experiment	Dispersant Blend[2] D1	: D2	Dispersion (# of inversions)	Suspension[1] initial	Suspension 2 weeks @ 54°C	Attrition resistance (%)
A	0	12	6	88	60	19
B	2	10	12	92	90	8
C	4	8	12	90	93	5
D	6	6	14	87	87	5
E	8	4	19	82	82	4
F	10	2	incomplete	65	65	2
G	12	0	incomplete	55	58	4

[1] CIPAC MT.15.5
[2] D1= low degree sulfonation Kraft lignin
D2= high degree sulfonation Kraft lignin

Experiment A showed a relatively high initial suspension. Shelf life instability was observed in this case after storage. As soon as small amounts of low degree of sulfonation dispersant (D1) are present in the formulation a plateau region of shelf life stable compositions can be detected.

Formulation A turned out to be relatively soft with relatively high attrition resistance (19%) and high dispersibility (6 inversions). Both attrition and dispersibility form a performance plateau region in the neighborhood of Experiments B and C. Granule hardness increases as soon as portions of low degree sulfonation dispersant are present in the blends. (Westvaco internal method. Available upon request).

The same seems to be the case for suspension stability. Storage stability improves as well when fractions of low degree sulfonation dispersant are present.

As can be seen from Table 3, Experiments B and C, the lowest yield values correlate well with higher extrusion throughput rates.

The proportion of elastic and viscous components in the various experiments was determined by creep / recovery test using a Haake RS 150 controlled stress rheometer. Attempts to determine the linear viscous elastic range were not successful with this series. Therefore the optimal shear stress value to be applied for creep/recovery tests had to be determined experimentally. To do so, a series of increasing stress values were selected observing the creep/recovery response. According to the smoothness and the regression value of the creep curve (r= >0.85) a lower and a higher stress region can be defined representing the lower and higher end of the deformability of the elastic string elements in the paste. Then a mid range stress value is selected from this region for the final creep measurements.

Results indicate that the highest elasticity values are found in the plateau formed by Experiments B and C.

Table 3 - *Influence of dispersant molecular weight and degree of sulfonation on paste rheology and extruder throughput rates.*

Experiment	Dispersant blend D1 :	D2	Yield Point[1] (Pa)	Extruder throughput rate (g/h)	Elastic/ viscous proportion (%)
A	0	12	645	140	~100% viscous
B	2	10	750	280	30/70
C	4	8	745	305	45/55
D	6	6	925	275	25/75
E	8	4	1180	180	viscous, hard
F	10	2	1690	120	viscous, hard
G	12	0	2990	80	dilatant, hard

[1] Yield point by deformation

As a conclusion to this series of experiments formulation B is selected for further investigation.
(Dispersant ratio D1: D 2 = 1:5)

The range of water requirement for extrusion resulted in:
15 to 17 % water (mean: 16%) determined by torque rheometry, and
14 to17.5 % water (mean: 15.8%) determined by yield by deformation

In case a torque rheometer is not available for research a stress rheometer can be used instead. Yield points of the extruder mass as a function of increasing water content using deformation methods are recorded. If the moistened powder mass is sufficiently deformable (G'> 20%) a plateau region similar to those observed by torque measurements can be observed. This is demonstrated in Figure 3.

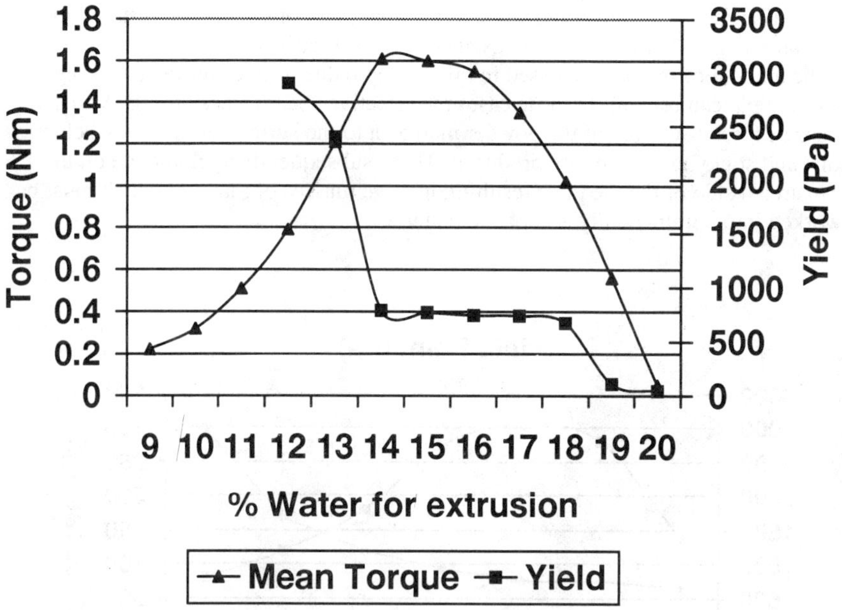

Figure 3 - *Mean torque plateau vs. yield by deformation.*

This diagram demonstrates the feasibility applying torque rheometry and yield measurements by deformation in order to determine the water requirement of feedstock pastes for low-pressure extrusion.

According to Leuenberger (Leuenberger and Bier 1990) the range of the possible water additions within the determined plateau region can be used to adjust the attrition resistance of the processed granule without negatively influencing most of the other physical parameters.

Table 4 shows the relationship of suspension, attrition resistance and rework within the limits of the pre-determined plateau range.

Table 4 - *Plateau Region: Paste water content amount vs. attrition and suspension.*

% Water (at plateau)	Attrition resistance (%)	Suspension	Rework (%)
14	15.8	89	5
15	12.0	92	5
16	8.5	92	9
17	8.0	91	8
18	8.0	90	21

As can be seen from these data, suspension is not affected when working within the limits of the defined plateau. At the lower end of the plateau formulations with less attrition resistance are noted while at the upper end of the plateau resulted in formulations with better attrition resistance are observed.

The amount of rework increased in this example due to overwetting (working above the plateau region). In those cases the effective shear forces during extrusion squeeze excess water out of the powder matrix onto the surface of the feedstock mass. Moist and sticky extrudates are produced. These subsequently agglomerate on the downstream side of the die plate resulting in large masses of material which must be reworked into commercially acceptable material.

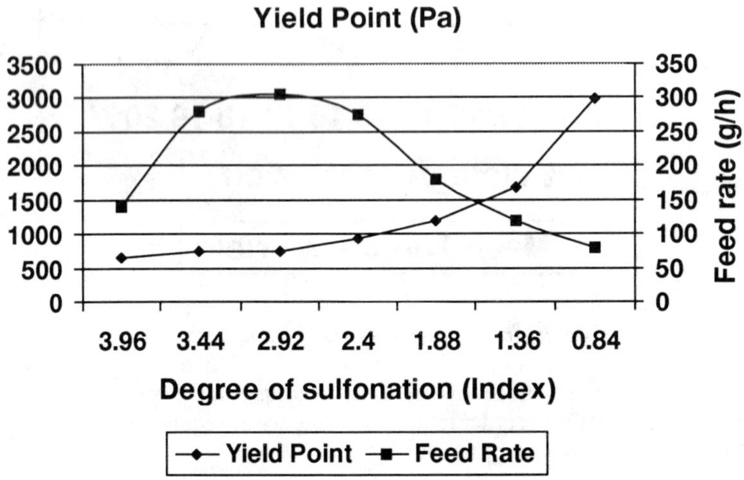

Figure 4 - *Influence of the degree of sulfonation on paste yield value and extruder feedstock rate.*

Reviewing the entire experimental design, it can be observed that extruder throughput rates increase, as small amounts of low degree of sulfonation dispersant

(D1) are present in the blends. This is consistent with the lower yield values recorded for experiments B, C and D.

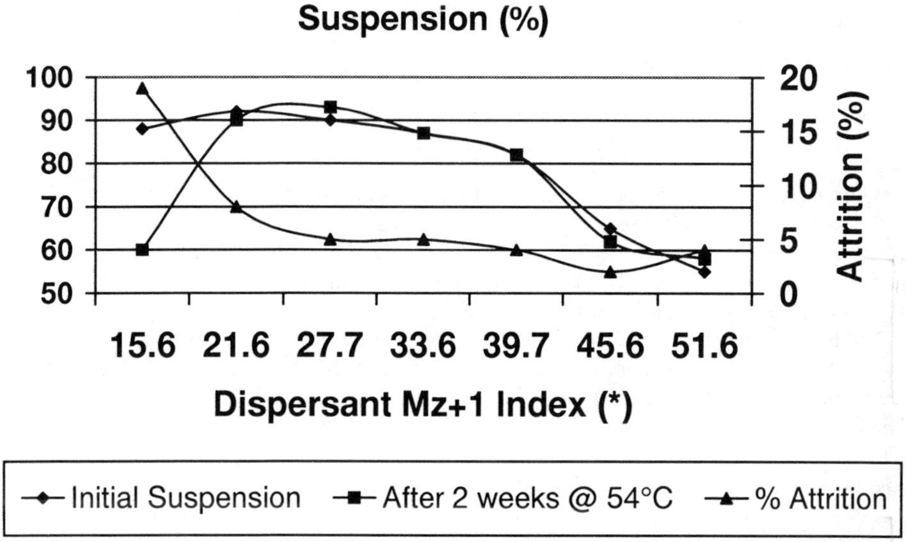

Figure 5- *Molecular weight effects: influence of dispersant Mz+1 average molecular weight on Suspension and attrition resistance.*

(*) = Mz+1 index represents the linear increase oft the degree of sulfonation in the final dispersant blend calculated by company internal methods. The Mz+1 value as such represents the high molecular weight end of a molecular weight distribution.

When comparing the influence of the dispersant Mz+1 average molecular weight on suspension and attrition resistance it becomes apparent that the often noted performance platform region around experiments B, C and D applies in this correlation as well. Storage stable suspensions with reasonable attrition resistance are present across this plateau.

In addition to torque rheometry and rheology measurements, pore size distributions determined by Mercury porosimetry also contribute to the emerging model of the extruded WDG.

Table 5 - *Pore size distribution vs. water for extrusion.*

Pore Parameter	14% Water	17% Water
Total pore area (m^2/g)	5.07	17.6
Median pore diameter (μ)	60.41	0.059
Average pore diameter (μ)	0.497	0.022
Porosity (%)	56.0	15.7

Comparing the low end of the plateau of paste water requirement using 14% water to the high end of the plateau using 17% water, it can be seen that higher water additions generally result in smaller pores and consequently lower porosity and surface area. The pore size differences within water requirement plateaus are not always as clear-cut as in this example, however.

The relative proximity of mean torque and mean amplitude measured by torque rheometry suggests a good and strong binder / substrate interaction, which would favor harder granules and subsequently incomplete dispersion at higher water levels. The fact that this does not occur can be attributed to apparently weak binding at a 1:5 (D 1: D 2) dispersant ratio, to the rapid dissolution of the binder layer, or both. (Benbow J.J. and Lawson T.A. 1989)

As water is sprayed onto the powder mixture part or all of the dispersant dissolves and forms a highly viscous solution. Depending on the amount of water present in the paste, the flow characteristics of such dispersant solutions would vary. This offers another window for additional formulation quality parameters. (Blackburn S. and Boehm H. 1993)

High viscosity surfactant liquids would flow more slowly into the pores and generally create larger pores or large pockets of concentrated surfactant within the pore structure, while lower viscosity surfactant liquids tend to form and fill smaller pores with a smaller pore distribution spectrum. When the formulation is added to water and diluted, larger openings on the granule surface permit rapid water access, thereby dissolving the surfactant in the pore structure and granule disintegration occurs.

Our model study involved the following surfactant solutions

Components: Dispersants (D 1: D 2) plus wetting agent = a total of 14%

Solution 1: @ 14% water results in a solution with 50% solids.

Solution 2: @ 17% water results in a solution with 45% solids.

Both solutions were analyzed rheologically with the following results

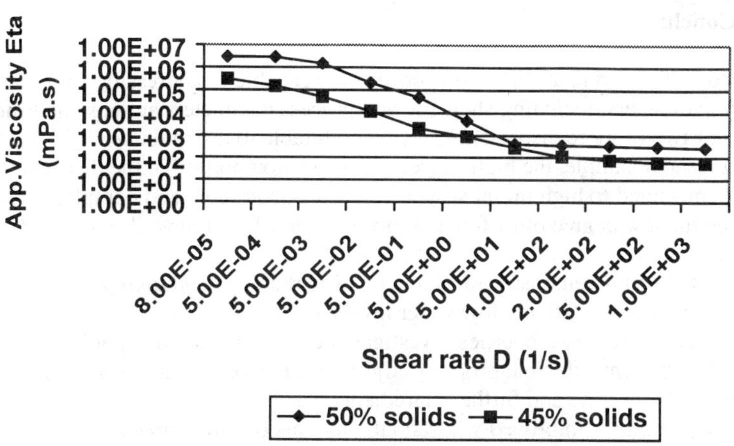

Figure 6 - *Low shear rheology of dispersant solutions.*

As can be seen from Figure 6, the higher water level of 17% produced a much more fluid surfactant rheology than expected. This is further substantiated in Table 6.

Table 6 - *Other rheological properties of the dispersant solutions*

Rheological Parameter	14% Water = 50% solids	17% water = 45% solids
Yield stress (Pa) Casson	3.2	0.094
Zero shear viscosity (Pa.s)	9 454	12 650
Equilibrium compliance (1/Pa)	0.002	0.002
Elastic recovery time (s)	124	24
Elastic /viscous ratio	1:1	1:5
Elastic recovery time (s)	124	24

Yield points are low for both systems indicating that a sufficient liquid flow is guaranteed in a shear environment (extruder).

The fact that the zero shear viscosity of the more dilute system with 17% water is higher than the zero shear viscosity of the more concentrated solution containing14% water can be explained in the following ways:

1. The higher water amount causes better flow at higher shear forces but allows the matrix to swell or to expand more when no shear is applied.
2. Viscous elements are more predominantly present in the system containing 17% water than they are in the system with 14% water only.

General Conclusions

1. Selected sulfonated Kraft lignin dispersants are capable of addressing formulation quality parameters, including shelf life and process parameters at the same time.
2. Blends of two or more dispersants may be preferable to a single one.
3. In the present example, the high degree of sulfonation portion in the dispersant blend contributed to high initial suspension levels, and rapid and complete break up while the low degree of sulfonation portion contributed towards enhanced shelf life stability.
4. The selected blend demonstrated excellent paste rheology and high paste throughput rates at relatively low water levels.
5. Several of the physical properties investigated exhibit plateau like performance ranges which permit flexible property adjustment for formulation design and scale up without the need to add further ingredients.
6. Shelf life stability is improved when small amounts of low degree of sulfonation Kraft lignin are blended into the formulation.
7. Elasticity (G') and elastic recovery (die expansion) reach ideal levels in correlation with extruder feedstock rates.
8. Paste rheology and extruder feed stock throughput rates can be influenced and optimized when selecting the ideal lignin chemistry for the system and by adjustment of the dispersant molecular weight (Mz+1 average) and the degree of sulfonation of the dispersant.

Acknowledgment

We would like to thank *Mr. Timothy M. Martin (FMC Agricultural Group)* for letting us use the Caleva torque rheometer and for his most valuable contributions during our experimental work.

References

Benbow, J. J., and Lawson, T. A., 1989, "Prediction of Paste Extrusion Pressure," *Ceramic Bulletin*, Vol. 68, No. 10, pp. 1821-1824.

Blackburn, S., and Boehm, H., May 1993, "The Influence of Powder Packing on Paste Extrusion Behavior," *Trans I Chem E*, Vol. 71, Part A.

Chatlapalli, R. and Bhagwan, D. R., 1998, "Rheological Characterization of Diltiazem HCL/Cellulose Wet Masses Using Mixer Torque Rheometer," *International Journal of Pharmaceutics*, 175, pp. 47-59.

Delli Colli, H. T., and Doering, I., 1999, "The Physical Chemistry of WDG's. I. Porosity and Surfactant Dissolution," *Pesticide and Application Systems: Twentieth Volume*, ASTM STP 1400, J.C. Clover, Ed., American Society for Testing and Materials, West Conshohocken, PA, 2001.

Leuenberger, H., and Bier, H. P., June 1990, "Theory of the Granulating Liquid Requirement in the Conventional Granulation Process," *Pharmaceutical Technology,* pp. 61-68.

Parker, M. D., and Rowe, R. C., 1990, "A Method for Quantifying the Consistency of Wet Granulation," *Pharmaceutical Technology International*, 2, pp. 50-62.

Parker, M. D., and York, P., 1990, "Binder-Substrate Interactions in Wet Granulation. 1: The Effect of Binder Characteristics," *International Pharmaceutical,* 64, pp. 207-216.

Alan J. Stern[1] and James L. Hazen[2]

Design and Performance of Crop Oil Concentrate Emulsifiers

Reference: Stern, A. J. and Hazen, J. L., **"Design and Performance of Crop Oil Concentrate Emulsifiers,"** *Pesticide Formulations and Application Systems: A New Century for Agricultural Formulations, Twenty First Volume, ASTM STP 1414,* J. C. Mueninghoff, A. K. Viets, and R. A. Downer, Eds., American Society for Testing and Materials, West Conshohocken, PA, 2001.

Abstract: Experimental design was used to study the interactions of selected low cost emulsifier components and thus identify a crop oil concentrate emulsifier blend having optimum performance and low cost. For this study, the mixture variables included: nonylphenol + 6 EO, nonylphenol + 10 EO, tall oil fatty aid (TOFA), and paraffinic petroleum oil. Water, although not an independent variable, was included as a coupling agent. The responses measured were number of phases, emulsion stability in soft water, emulsion stability in hard water, and emulsion stability in 2.5% ammonium sulfate solution, all at ambient temperature.

Using experimental design software, it is possible to visualize the interrelationships of the four components for each test condition, making it possible to predict the performance of any composition within the design space. After analyzing the response surface contour maps of the four-dimensional design space, we find that it is possible to formulate a crop oil concentrate that is single phase, offers very good emulsion performance in soft and hard water, and performs moderately well when tank mixed with ammonium sulfate solution. Coupling the performance data with a simple cost calculation, the formulator can select a composition that maximizes performance while at the same time minimizing cost.

Keywords: Crop oil concentrate, experimental design, response surface, contour, emulsifier, coupling agent

Introduction

Crop oil concentrate (COC) adjuvants have been in use for many years. These adjuvants typically contain 83 % petroleum oil and 17 % emulsifier (Thompson 1998). Over the years, there has been intense competition among emulsifier suppliers that has led to a downward pressure on the market price for the emulsifier component of the COC. Prices have been further eroded by a decrease in grower demand for COC products as agronomic practices have changed. Recently, as end users have

---

[1] Scientist, Agricultural Specialties, Crompton Corporation, 5777 Frantz Road, Dublin, OH, 43017.

[2] Global R&D manager, Agricultural Specialties, Industrial Specialties 5777 Frantz Road, Dublin, OH, 43017.

demanded greater flexibility of use of crop oil concentrates, suppliers have responded with premium "value added" products designed for better performance under more difficult use conditions. Compatibility with liquid nitrogen fertilizer or ammonium sulfate-containing spray solutions are typical of this type of enhanced COC.

Responding to these pressures, emulsifier suppliers have sought to design effective, yet low cost emulsifier blends. "Effective" in this context means physical compatibility with the petroleum oil for the shelf life of the product, as well as good end use performance in the spray tank. As will be shown below, the results of the present study could be used to identify optimum compositions of a COC having more or less emulsifier than the traditional 83/17 blend. A blend with good performance and less than 17% emulsifier would benefit from a lower cost structure and would therefore have an advantage in a highly competitive marketplace. Increasing the ratio of emulsifier to oil might be expected to enhance the biological response of the spray mixture through better wetting and spreading of the spray on the target organism.

To achieve these goals, experimental design was used to study the interactions of selected low cost emulsifier components and thus identify a crop oil concentrate emulsifier blend meeting the desired performance criteria. For this study, the variables included: nonylphenol + 6 EO (NP-6), nonylphenol + 10 EO (NP-10), tall oil fatty acid (TOFA), and paraffinic petroleum oil. By including "oil" as a variable, we are not limited to the traditional 83/17 type COC. In this way, a better understanding of the interrelationships between oil, coupling agents, and surfactants is gained. Water, although not an independent variable, was included as a coupling agent. The responses measured were number of phases, emulsion stability in soft water, emulsion stability in hard water, and emulsion stability in 2.5% ammonium sulfate (AMS) solution at ambient temperature.

Experimental

ECHIP[3] was used to create the design and calculate the response surfaces (Wilson 1997). A four-dimensional partial cubic design was established in which the following limits were placed on the variables: paraffinic oil 80-90%, TOFA 6-11%, NP-6 2-6%, NP-10 2-6%. The design has a G-efficiency of 59.2% and contains 30 unique, and 5 replicate points. G-efficiency measures how well the design covers the entire four-dimensional design space. For best results, a G-efficiency greater than 50% is desirable. Because a small amount of water is often useful as a coupling agent in COC formulations, the TOFA, NP-6, and NP-10 each were diluted with 8% water before use. The TOFA/water mixture was kept agitated during use to prevent separation.

Twenty grams of each of the 35 trial mixtures was prepared in one ounce glass vials. The vials were thoroughly mixed and allowed to stand at about 22 °C overnight. The samples were then visually evaluated for number of phases.

Emulsions were prepared at 22 to 24 °C by adding one mL of the test formulation to 99 mL of carrier. Carriers were 2.5% AMS[4], 34 ppm hard water and 1000 ppm hard water, prepared in accordance with ASTM Method for Emulsification Characteristics of Pesticide Emulsifiable Concentrates (E 1116). The tubes were inverted ten times then

[3] ECHIP, Inc., 724 Yorklyn Rd., Hockessin, DE, 19707, version 6.4.1.

[4] Prepared from reagent grade diammonium sulfate and 342 ppm hard water.

allowed to stand at ambient temperature. After one hour, the emulsions in 2.5% AMS were evaluated. After two hours, the emulsions in soft and hard water were evaluated. In the figures, a value of "10" indicates a near perfect emulsion, showing no signs of separating cream. A value of "1" indicates a very poor emulsion that has almost completely separated in a cream or oil phase at the air/water interface. The emulsion ratings and number of phases of each test mixture were used by ECHIP to derive the equations describing the response surfaces (and thus the contour diagrams.)

Results and Discussion

Interpreting the Contour Diagrams

To visualize the interelationships between the four components in our four-dimensional design space, it becomes necessary to hold one mixture variable constant so that the remaining three may be plotted on a conventional ternary diagram. This allows visualization of a three-dimensional "slice" through the four-dimensional space. Throughout this text, we have elected to show slices in which the oil level is held constant. For each contour diagram, the oil level can be found to the upper right of each plot. For example in Figure 1, "Oil = 0.83" means that in the left ternary diagram, oil is held constant at 83%. The three sides of each ternary diagram are labelled with the component (variable) and percent. Also, below many of the contour diagrams is the composition, predicted response value, and the 95% confidence limits for that predicted value for the point indicated by the cursor on the diagram. The cursor appears as a small diamond with a cross through it.

Furthermore, some figures show contours for regions outside investigated design space. When this is the case, the investigated design space is the area within the bold polygons. Although the software will give predictions outside this region, they usually have poor reliability.

Phase Behavior

Phase data was used to generate contour plots that are useful in understanding the phase behavior of the COC system. Beginning with 83% oil, the system moves closer to two phases as the amount of TOFA decreases (Figure 1). This is expected since the primary function of TOFA in this blend is to couple the oil and emulsifier together. As one moves across the response surface toward higher oil levels (the fourth dimension) by reducing emulsifier and coupling agent levels, the single-phase region shrinks, again as expected. Thus, with less oil in the blend, a larger one-phase region is available in which to optimize emulsification. As a result of drawing contour diagrams at different oil levels, it becomes impossible to draw the diagrams to the same scale (Figure 1).

Soft and Hard Water Emulsion responses

Next, examine the contour diagrams for emulsion performance in soft and hard water. As before, we will look at slices through the four-dimensional response surface in which the "% oil" is held constant. With "oil" held at 81%, there are similarities between the "hard" and "soft" water responses. Performance tends to increase in both waters as the ratio of TOFA/NP-6/NP-10 approaches 46:30:24. Interestingly, as the oil level is increased a few percent to 84%, the response surfaces and the optimum ratio of TOFA/NP-6/NP-10 change significantly. At this oil level, the optimum ratio is more heavily weighted in TOFA; a ratio of 50:32:19 is best. This more lipophilic blend of TOFA, NP-6, and NP-10 is required to match the greater lipophilicity of the system containing more oil.

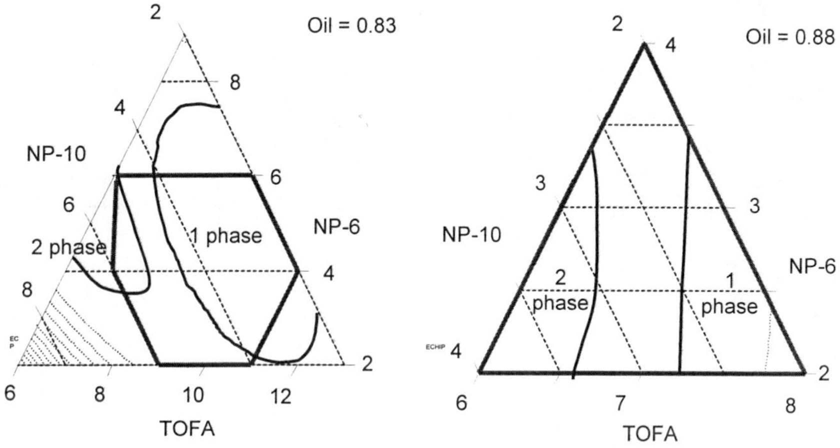

Figure 1- *System phase behavior at 83% and 88% oil.*

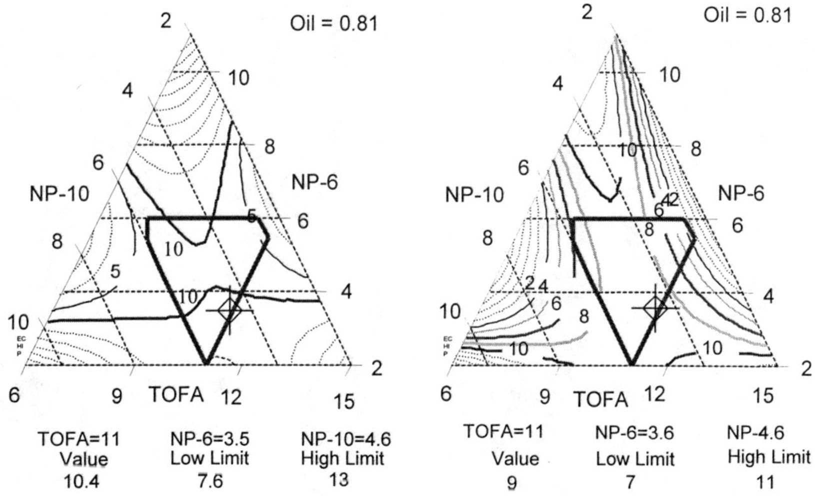

Figure 2-*Response surfaces: emulsions in soft and hard water at 81% oil at two hours.*

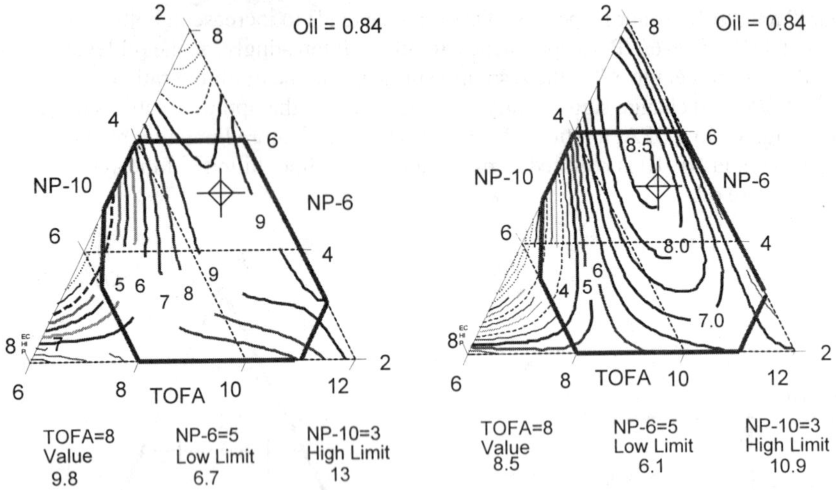

Figure 3 *Response surfaces: emulsions in soft and hard water at 84% oil at two hours.*

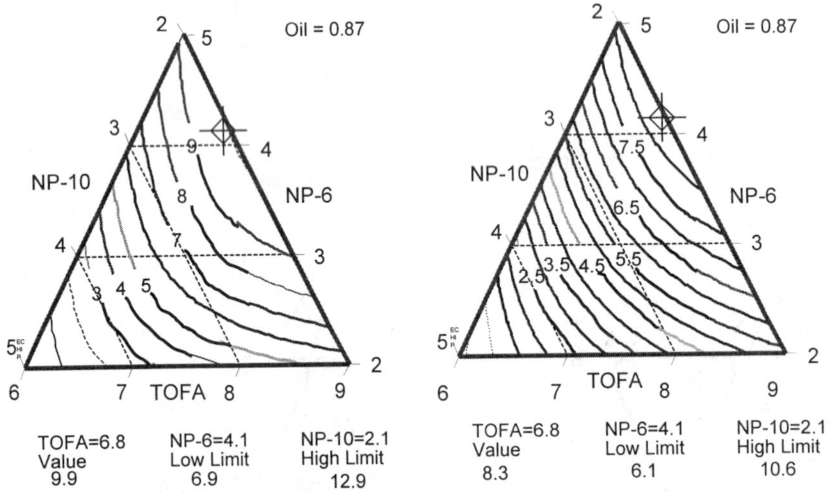

Figure 4-*Response surfaces: emulsions in soft and hard water at 87% oil at two hours.*

Table 1- *Optimum ratio of TOFA/NP-6/NP-10 for different oil levels.*

OIL LEVEL	TOFA	NP-6	NP-10
81%	46	30	24
84%	50	32	19
87%	53	32	15

As the oil level is pushed further to 87%, the overall emulsion quality begins to decline, and the optimum ratio of the three variables continues to move to a more lipophilic composition, such as 53:32:15. This is summarized below (Table 1 and Figures 2 to 4).

Emulsions in 2.5% Ammonium Sulfate

As mentioned earlier, many users are demanding value-added, enhanced performance COC's. Tank mix compatibility with multiple agchemical mixtures and compatibility with dilute fertilizer solutions, including ammonium sulfate, are examples of desired attributes. To address this need within the scope of the low-cost emulsifier system, emulsion performance in 2.5% ammonium sulfate solution was included as one of the experimental design responses. The results of the experimental design in this situation are more difficult to interpret. As before, it is found that at higher emulsifier levels, i.e. 15-20%, the system prefers an emulsifier blend high in NP-10, and low (relative to the NPE's) in TOFA. However, starting at about 15% emulsifier and more prominently as the emulsifier load is decreased, a new maximum in the response surface emerges in a region of higher TOFA and NP-6. Part of the explanation for this shift lies in the lower HLB of the COC system containing a greater amount of oil. One conclusion, however, is indisputable: as can be seen in the emulsion ratings, more surfactant is better when the COC is used in highly ionic solutions such as ammonium sulfate (Figure 5, Table 2).

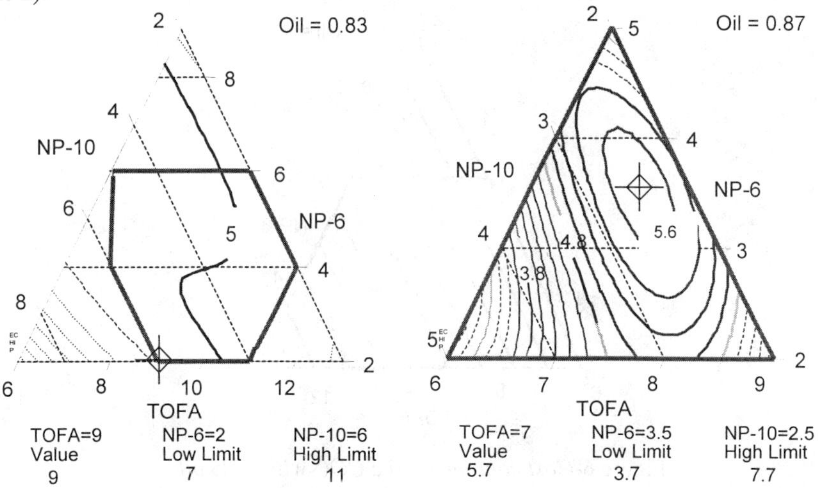

Figure 5- E*mulsions in 2.5% ammonium sulfate at 83 and 87% oil.*

Table 2- *Optimum emulsifier blends for emulsions in 2.5% ammonium sulfate.*

OIL, %	TOFA	NP-6	NP-10	PREDICTED PERFORMANCE
81	57	11	32	9
83	53	12	35	9
85	47	14	39	7
85	41	38	21	6
87	54	27	19	6

The Role of Product Cost

Having thoroughly investigated the physical performance of the blends within the design space one can easily find the optimum composition for any given test condition, as well as the best composition averaged over all the responses. Now consider cost. Does the optimum composition have an acceptable cost? Must performance be sacrificed in order to compete in the marketplace on price? To better understand how cost varies with composition, cost contours have been plotted on the same type of ternary diagrams used to show emulsion performance. To calculate the contours, the cost of the four components was arbitrarily set as follows: petroleum oil, \$0.15/pound, TOFA, \$0.26/pound, NP-6, \$0.96/pound, and NP-10, \$0.95/pound. The contours run parallel to the NP-10 axis, when the "oil" level is held constant (Figure 6). It is not surprising that the COC cost is reduced as the oil level and/or TOFA level is increased at the expense of the nonylphenol ethoxylates. In Figure 6, the contour values represent the cost of the COC formulation in cents per pound.

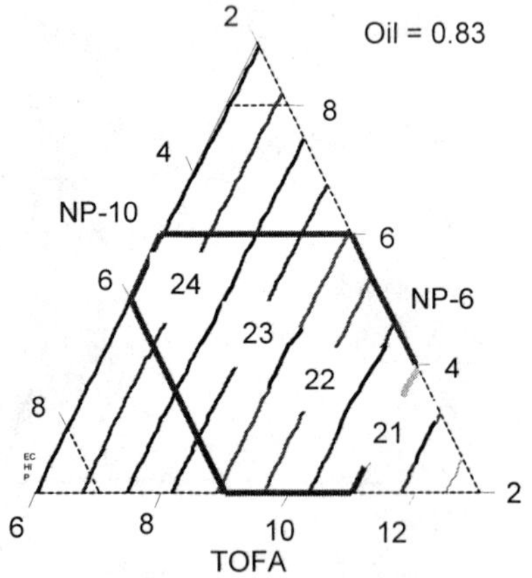

Figure 6-*Cost contour for the COC with 83% oil.*

Combined Response Surface-Optimization of Cost and Performance

One of the advantages of computer-driven experimental designs is the ability to optimize several response functions simultaneously. Thus, to find the overall best composition for the blend of petroleum oil, TOFA, NP-6, and NP-10, a combined response contour diagram has been created (Figure 7), in which the oil level is 83%. The diagram shows a ridge of high performance running from the left of the apex toward the lower right corner. When this diagram is overlaid on the cost contour diagram, it becomes apparent that our best choice for low cost and high performance in a traditional 83/17 COC would correspond to a point on the ridge, but as far to the lower right as possible. Thus, the experimental design predicts that a blend containing[5] 83% oil, 10.2% TOFA, 3.8% NP-6, and 3.0% NP-10 ought to be the best possible compromise between performance and cost under the test conditions described herein. This composition was tested. It was found to be one phase and fully satisfactory in soft and hard water, but weak in 2.5% AMS solution. This can be attributed to the calculation of the combined response, which puts equal weight on the four responses (phases, soft, hard, AMS). The first three of these coincide nicely, while the best AMS performance is found in a much different location in the design space. Chemically speaking, this relates to the need for a more hydrophilic emulsifier in the presence of AMS.

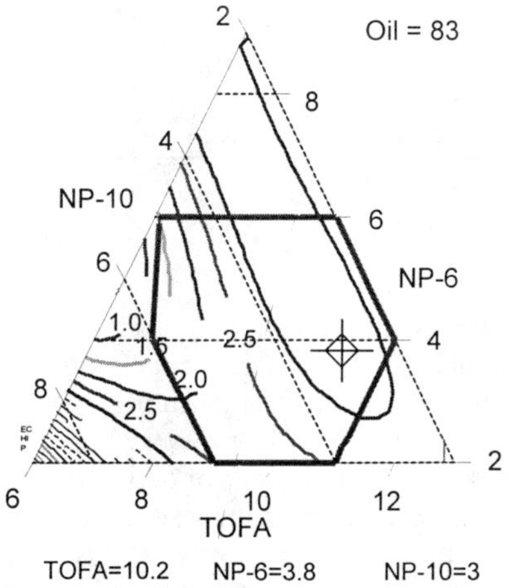

Figure 7-*Combined response contour diagram showing the point of best cost and performance.*

[5] Although not explicitly shown, note that a small amount of water, about 1.36% in this case, is included as a necessary coupling agent. See the experimental section for details.

Review

This study has shown that as the level of oil in the COC is increased or decreased from the traditional 83%, the best ratio of TOFA, NP-6 and NP-10 changes in a predictable way. The level of TOFA, the primary coupling agent, is critical to maintaining a one-phase COC product. In standard test waters, the HLB of the optimum blend drops as the amount of oil in the COC is increased from 80 to 90%. Emulsion performance in the presence of 2.5% AMS is complex in that at high emulsifier levels, there appear to be two "good" regions with different compositions. At low emulsifier levels, a single, well-defined sweet spot appears. Cost is not a physical "response," but it is a necessary consideration in any product development. Cost as a function of composition also can be visualized on a ternary contour diagram. For the compositions under study, evenly spaced cost "contours" run nearly parallel to the NP-10 axis. By combining the phase diagram with the various emulsion performance contour diagrams, and taking into consideration raw material costs, an optimum emulsifier blend for the traditional 83/17 COC adjuvant is presented (Table 3).

Table 3-Optimized 83/17 COC composition.

Component	Percent
Petroleum Oil	83
TOFA	9.38
NP-6	3.5
NP-10	2.76
Water	1.36

Summary

Crop oil concentrates, an important segment of the herbicide adjuvant market, must be formulated to emulsify well under many different conditions and be cost effective. With the use of experimental design, the interrelationships of the components of a simple crop oil concentrate system were examined. The response surfaces resulting from the design lead the formulator quickly to compositions of optimum phase stability, soft water or hard water emulsion performance, and emulsion performance in ionic solution. Many other performance parameters could be evaluated as well. Also, the issue of raw material cost has been considered. Again, a contour diagram was found useful for this analysis. By combining the four "physical" responses, a "global optimum" composition was picked in such a way that performance was maximized and cost was minimized.

References

Thompson, L., 1998, *A Guide to Agricultural Spray Adjuvants used in the United States*, 5th edition, Thompson Publications, Fresno, CA pp 81.

Wilson, P., 1997, "Use of Multi-Variable Analysis in the Development of a Medical Diagnostic Reagent from Cultured Human Cells," *Biotechnology International,* Universal Medical Press, Inc. San Fransisco, CA, 94107, pp 95-104.

SESSION IV: Formulation Ingredients I

Richard Herbert[1]

Novel Polymeric Dispersants for Water Dispersible Granules

Reference: Herbert, R., **"Novel Polymeric Dispersants for Water Dispersible Granules,"** *Pesticide Formulations and Application Systems: A New Century for Agricultural Formulations, Twenty First Volume,* ASTM STP *1414,* J. C. Mueninghoff, A. K. Viets, and R. A. Downer, Eds., American Society for Testing Materials, West Conshohocken, PA, 2001.

Abstract: High concentration extruded water dispersible granules (WG) can be processed using styrene acrylic and modified styrene acrylic dispersants at formulation concentrations usually found in wettable powder formulations: 3-7 % (w/w). These dispersants have found application with a wide range of active ingredients at concentrations ranging from 75% to >90% (w/w) of technical grade active ingredient. Extruded granule formulations containing these dispersants demonstrate excellent granule disintegration and physical suspension upon dilution both initially and upon extended storage at elevated temperatures. Experimental design techniques have also been identified which dramatically shorten the product development time frame.

Keywords: Dispersant, water dispersible granule, styrene acrylic polymer, experimental design, screening method, product development

Introduction

With agricultural product development, the formulation chemist must view the world from two perspectives: cost and performance. This often places him/her in the position of trying to serve two masters at the same time. In order to minimize formulation cost, it is desired to develop the highest formulation concentration possible; however, as the concentration increases, there is often a trade-off in processing and/or application performance. Such is the case with high concentration extruded WG development.

As the active ingredient concentration of extruded granule formulations increases, the choice and concentration of excipient components becomes increasingly critical since the space available within the formulation to meet their functional requirement becomes less and less. In other words, the choice of wetting agent, dispersant and diluent must be from the most efficient (lowest application rate requirement) chemistries available.

This research has identified a unique class of chemistry, styrene acrylic polymers, which function as dispersants in the extruded WG formulations at rates significantly lower than the currently available sodium lignin sulfonate and sodium naphthalene sulfonate formaldehyde condensate chemistries. These rates have been experimentally determined to be as low as 3% (w/w) on a solids basis. The styrene acrylic dispersant

[1] Development Associate, Uniqema, 900 Uniqema Boulevard, New Castle, DE, 19720

polymers are identified in commerce as ATLOX®[2] Metasperse 100L and ATLOX Metasperse 150S for the unmodified form and as ATLOX Metasperse 500L and ATLOX Metasperse 550S for the modified form of the polymer. In the trademark name, the "L" infers liquid and "S" infers solid physical state. Liquid forms are approximately 30-35% of the styrene acrylic polymer in water while solid forms are greater than 90% polymer.

This research has also identified an approach that allows the formulation chemist a means to obtain the full benefit of the new chemistry. The approach provides the chemist not only with a 'map' for product development, using styrene acrylic polymer dispersants, but also with a framework in which their performance benefits versus industry standards can be easily determined. It is not a new approach but rather a new application of an old tool known as experimental design.

Materials and Methods

The extruded granule technology development process involves the following:

1. Definition of an extruded granule,
2. Establishment of criteria of performance,
3. Determination of test methods,
4. Formulation processing,
5. Comparison of styrene acrylic polymer dispersants with industry standards, and
6. Application with alternative technicals

Definition of an Extruded Granule

An extruded granule is a delivery system for a wettable powder, which is itself a dry physical blend of formulation components. The extruded granule, however, is unique in that it addresses concerns with dust exposure at the time of application. By defining an extruded granule in this manner, the implication is that the granule must demonstrate equivalent physical suspension to the wettable powder and disintegration within the same time frame as the wettable powder wets.

Establishing Criteria of Performance

Performance criteria for extruded granules focuses upon the properties of wetting/disintegration and physical suspension upon dilution. These criteria were established from the perspective of acceptable wettable powder formulation performance and were defined as follows:

- Disintegration = >95% within three (3) minutes, and
- Physical Suspension = >90% after thirty (30) minutes standing.

Determining Test Methods

Equipment used for milling and mixing wettable powders and for manufacturing (by extrusion) water dispersible granules were a Retsch ZM 100 centrifugal mill, an LCI KAR-75 benchtop basket granulator, and a Denver Instruments TL-104 laboratory balance. Suspension and disintegration evaluations were performed in glass stoppered

[2] ATLOX® trademark of ICI Americas, Bridgewater, NJ

100 ml graduated cylinders. Time was kept using a standard digital stopwatch. The materials used as standard formulation components in all powders and granules were MORWET®[3] EFW as the wetting agent and kaolin clay (Barden AG-1) as the diluent.

Physical suspension expressed as suspension number was determined according to the following procedure:

1. Three samples of each powder or granule formulation is accurately weighed to one gram (± 0.0100 g).
2. Each sample is poured into a separate graduated cylinder containing 100 ml of either 342 ppm, 1000 ppm, or 2000 ppm hardness water.
3. The powders/granules are allowed to stand and wet for 5 minutes followed by mixing into the water phase (10 manual inversions) and again allowed to stand.
4. After standing for 5 minutes, each cylinder is remixed (10 inversions) and allowed to stand, this time for 30 minutes.
5. The top 90 ml (90% by volume) are siphoned off and the remainder collected on preweighed filter paper, dried and the weight determined.
6. Suspension number is calculated as $[1 - (\text{weight}_{\text{residue}} \div \text{weight}_{\text{sample}})] \cdot 111$.

Alternately, suspensibility values may be determined by ASTM Method E-1673[4] and expressed as percent suspensibility.

The physical suspension methodology was modified in order to focus more closely upon the degree to which the extruded granule disintegrated to its primary wettable powder particle size at the time of addition to the water. An arbitrary time frame of three (3) minutes was established. Granule disintegration was determined in the following manner:

1. One sample of granule formulation is accurately weighed to one gram (± 0.01 g).
2. The sample is poured into a graduated cylinder containing 100 ml of 342 ppm hardness water.
3. The granules are allowed to stand and wet for 3 minutes followed by mixing into the water phase (10 manual inversions).
4. After standing for 10 minutes, the top 90 ml (90% by volume) are siphoned off and the remainder collected on preweighed filter paper, dried and the weight determined.
5. Disintegration number is calculated as $[1 - (\text{weight}_{\text{residue}} \div \text{weight}_{\text{sample}})] \cdot 111$.

Formulation Processing

The purpose of experimental design is twofold. First, it identifies the initial extruded granule formulations to be processed, independent of the development chemist's biases, once the overall concentration of the components to be blended has been established. Second, it provides a means to monitor whether formulation changes positively or negatively affect performance according to agreed upon test methods and performance criteria.

---

[3] MORWET® trademark of Crompton Corporation, Greenwich, CT

[4] *Annual Book of ASTM Standards*, Vol. 11.05, ASTM, 100 Barr Harbor Drive, West Conshohocken, PA

As stated above, a basic wettable powder/extruded granule consists of a physical blend of:

- Technical,
- Wetting Agent,
- Dispersant, and
- Diluent.

It can therefore be visualized within the experimental design approach as a tetrahedron.

For a specific active ingredient concentration, extruded granule development can be viewed as a blending of wetting agent, dispersant and diluent among the remaining formulation percent (w/w); and can be visualized as a triangle, or a slice through the tetrahedron. While results of performance evaluation are often represented as a triangular phase diagram or contour map, the formulation triangle itself can be expressed as a numerical grid in which each number represents a specific ratio of formulation excipients as identified in Figure 1.

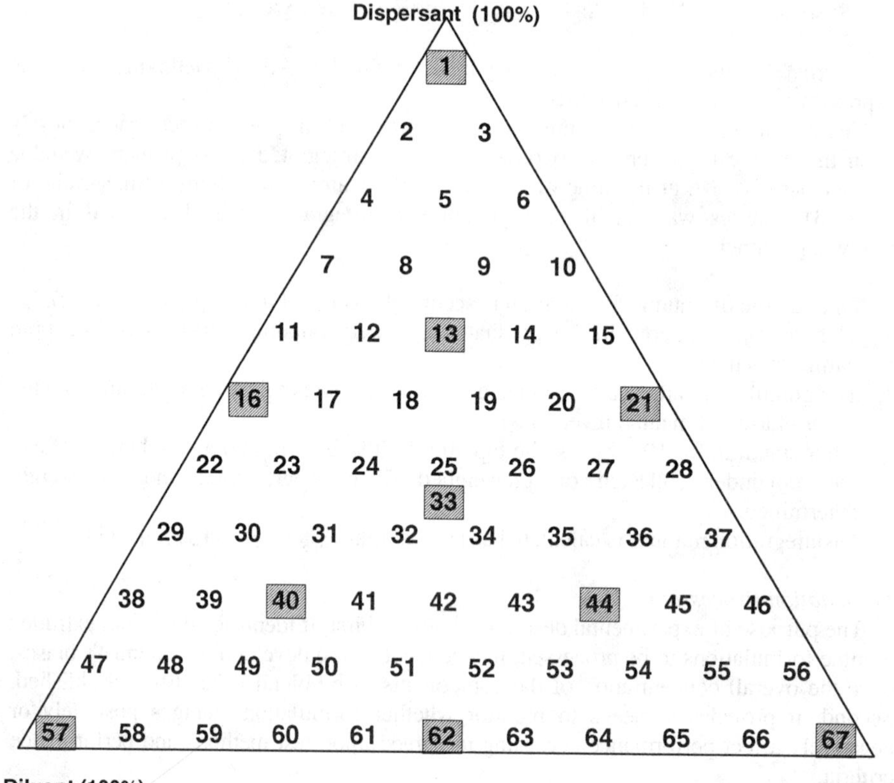

Figure 1. Formulation triangle represented as discrete points.

Using experimental design, the number of formulations to be processed can be significantly reduced through the identification of specific ratios which are representative of the formulas which surround it. These are referred to in this research as "The Primary Screening Points" and consist of each component alone, components combined in pairs, and as individual three-component blends. The standard ratios appearing in Table 2 below relate directly to the highlighted numbers appearing in Figure 1.

Table 1. Three-component blending primary screening points.

Screening Point No.	Formulation Excipient Ratios		
	Dispersant	Wetting Agent	Diluent
1	100%		
13	60%	20%	20%
16	50%		50%
21	50%	50%	
33	33%	33%	33%
40	20%	20%	60%
44	20%	60%	20%
57			100%
62		50%	50%
67		100%	

The above ratios are then converted to formulas for a given technical active ingredient at a desired active ingredient assay, processed as wettable powders, and evaluated for physical suspension and granule disintegration. The data is then plotted on the triangle and additional excipient ratios, based upon meeting acceptable performance criteria, are formulated and evaluated until the 'best' formulation is determined.

Comparison of styrene acrylic polymer dispersants with industry standards

As an example the development of carbaryl technical (97.5%) as a 75% WG was undertaken, comparing the performance of the styrene acrylic polymer dispersant to sodium lignosulfonate and sodium naphthalene sulfonate formaldehyde condensate dispersants in extruded granules.

As stated above, the process of comparative evaluation begins with the calculation of formulas to reflect the excipient ratios identified by the screening points listed above. It was decided to restrict blending on the triangle to a total of 15% (w/w) since, as a worst case, this would be divided between dispersant and wetting agent. In the best case, it would be shifted toward increasing concentrations of diluent.

For purposes of evaluation, then, the generic formulation identified in Table 2 was adopted. The formulas in Table 3 represent the specific inert ingredient compositions identified by the highlighted primary screening points on the triangle appearing in Figure 1.

Table 2. Model formulation with carbaryl.

Component	Total in Formulation (w/w)
Carbaryl Technical (97.5%)	75 %
Dispersant/Wetting Agent/Diluent Blend	15 %
Diluent	10 %

Table 3. Specific carbaryl 75 WG screening formulas.

Screening Point No.	Formulation Excipient Ratios		
	Dispersant	Wetting Agent	Diluent
1	15.0%		10.0%
13	9.0%	3.0%	13.0%
16	7.5%		17.5%
21	7.5%	7.5%	10.0%
33	5.0%	5.0%	15.0%
40	3.0%	3.0%	19.0%
44	3.0%	9.0%	13.0%
57			25.0%
62		7.5%	17.5%
67		15.0%	10.0%

These formulations were then processed and evaluated in the laboratory for physical suspension, both as wettable powders (Table 4) and as extruded granules (Table 5). Two key observations can be made from the data in Tables 4 and 5 above relevant to dispersant performance. First, physical suspension of the wettable powder may not be predictive of extruded granule performance. Second, a significant falloff in physical suspension property of the sodium lignosulfonate and sodium naphthalene sulfonate formaldehyde condensate dispersants at low concentrations is evident.

Table 4. Carbaryl 75WP physical suspension in 342 ppm hardness water.

Point Number	Dispersant Type			Comments
	Styrene Acrylic	Ligno-sulfonate	Naphthalene Condensate	
1	0	0	0	Non-Wetting
13	94	93	94	
16	0	0	0	Non-Wetting
21	94	93	96	
33	93	92	94	
40	90	91	92	
44	89	92	93	
57	0	0	0	Non-Wetting
62	0	0	0	Flocculated
67	0	0	0	Flocculated

Table 5. Carbaryl 75WG physical suspension in 342 ppm hardness water.

Point Number	Dispersant Type			Comments
	Styrene Acrylic	Ligno-sulfonate	Naphthalene Condensate	
1	0	0	0	Non-Disintegrating
13	94	94	94	
16	0	0	0	Non-Disintegrating
21	95	94	94	
33	93	60	93	
40	91	55	67	
44	91	91	89	
57	0	0	0	Non-Disintegrating
62	0	0	0	Non-Disintegrating
67	0	0	0	Non-Disintegrating

It appears that, until the minimum dispersant requirement is met, extruded granules will demonstrate a significant physical suspension loss upon granulation. This minimum dispersant requirement appears to be characteristic of the dispersant hydrophobe chemistry.

From this data, it immediately becomes obvious that wetting agent must be included into the carbaryl 75WG in order for the formulation to wet. Also, dispersant must be included into the carbaryl 75WG in order for the formulation to remain in suspension. Therefore, formulas represented by screening point numbers 1, 16, 57, 62, and 67 need not be processed in the future during technical or dispersant evaluation as an extruded granule.

Performance of both the sodium lignosulfonate and sodium naphthalene sulfonate formaldehyde condensate dispersants at 3% (w/w) concentration, strongly suggests that extruded granule physical suspension is a function of both wetting agent and dispersant concentrations (as observed with performance of point 40 versus point 44 for each dispersant).

Table 6. Carbaryl 75WG disintegration in 342 ppm hardness water.

Point Number	Dispersant Type			Comments
	Styrene Acrylic	Ligno-sulfonate	Naphthalene Condensate	
1	0	0	0	Non-Disintegrating
13	102	100	97	
16	0	0	0	Non-Disintegrating
21	100	96	54	
33	101	63	67	
40	97	67	91	
44	100	95	90	
57	0	0	0	Non-Disintegrating
62	0	0	0	Non-Disintegrating
67	0	0	0	Non-Disintegrating

The above carbaryl 75WG formulations were also evaluated for disintegration in 342 ppm hardness water according to the methodology delineated earlier. Results appear in Table 6. From Table 6, note that extruded granule disintegration is a function of surfactant concentration (wetting agent + dispersant) and that it varies with each dispersant chemistry. Overlaying the physical suspension and disintegration performances of the three dispersants on the triangle, it can also be seen that granule disintegration and physical suspension are two totally distinct physical properties which may or may not be related. For example, point 40 for the sodium naphthalene sulfonate formaldehyde condensate dispersant demonstrates very good physical suspension but only fair granule disintegration. The same point for the styrene acrylic dispersant demonstrates both excellent physical suspension and granule disintegration.

Using the above data to select a carbaryl 75WG formulation which demonstrates acceptable physical suspension and granule disintegration suggests that both the sodium lignosulfonate and sodium naphthalene sulfonate formaldehyde condensate dispersants require high surfactant (12-15% wetting agent + dispersant) concentrations. The styrene acrylic dispersant requirement, on the other hand, is approximately 6% (3% wetting agent + 3% dispersant).

The highlighted areas in Table 7 identify formulations which met the extruded granule criteria of performance for physical suspension (>90%) and granule disintegration (>95%).

Application with alternative technicals

Using the experimental design blending approach described above, it was possible to develop six (6) different solid technicals as 75% (w/w) technical WG using the styrene acrylic polymer dispersant at approximately 3-5% (w/w). Increasing the active ingredient concentration did not significantly affect the dispersant requirement. In addition, one liquid and two low melting solid technicals were formulated as extruded granules.

Table 7. Carbaryl 75WG physical suspension/disintegration in 342 ppm hardness water.

Point Number	**Dispersant Type**			**Comments**
	Styrene Acrylic	**Ligno-sulfonate**	**Naphthalene Condensate**	
1	0/0	0/0	0/0	Non-Disintegrating
13	94/102	94/100	94/97	
16	0/0	0/0	0/0	Non-Disintegrating
21	95/100	94/96	94/54	
33	93/101	60/63	93/67	
40	91/97	55/67	67/91	
44	91/100	91/95	89/90	
57	0/0	0/0	0/0	Non-Disintegrating
62	0/0	0/0	0/0	Non-Disintegrating
67	0/0	0/0	0/0	Non-Disintegrating

It was observed that four of the six solid technicals (diuron, carbaryl, chlorothalonil, and atrazine, all formulated as 75% technical WG) demonstrated acceptable performance directly substituted into the same formula (Table 8) One technical (iprodione) performed acceptably as a 75% technical WG with 3% (w/w) styrene acrylic polymer dispersant, however its wetting agent requirement was increased to 9.00% (w/w) under the conditions of this evaluation.

Table 8. Screening formulation for 75% technical WG.

Component	Total in Formulation (w/w)
Technical	75.0 %
Styrene acrylic polymer dispersant	3.0 %
Wetting agent	3.0 %
Diluent	19.0 %

With the resulting inert ingredient ratio selection for the technicals appearing in Figure 2, it appears that each technical has its own specific surfactant (styrene acrylic polymer dispersant + wetting agent) requirement depending upon technical surface properties: the more hydrophobic the technical surface, the greater the wetting agent requirement.

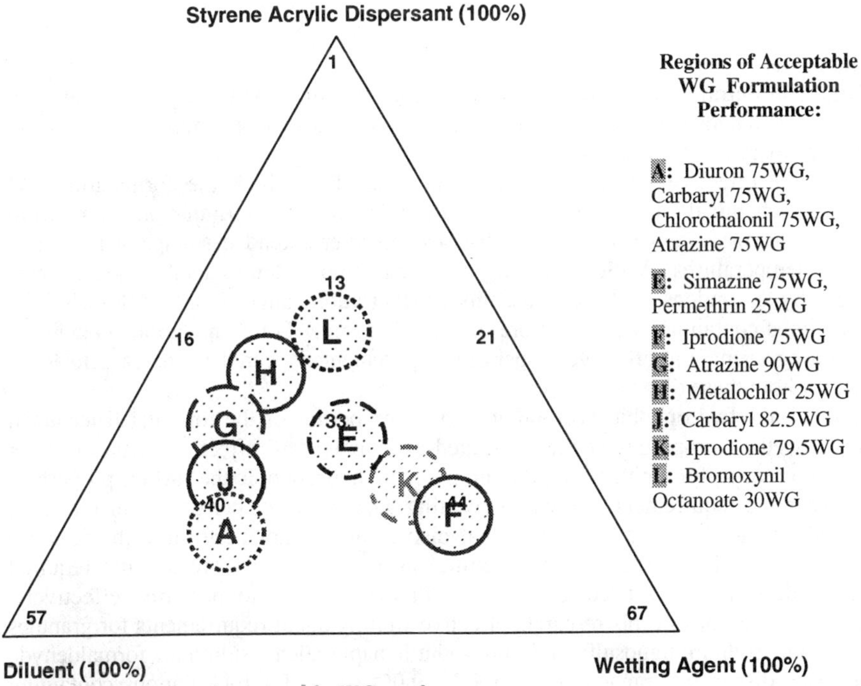

Figure 2. Regions of acceptable WG performance.

The implication of this observation may be that poor extruded granule performance, which in the past had been attributed to dispersant failure, may have actually been related to inadequate wetting agent in the formulation. Blending of dispersant, wetting agent and diluent within the framework of the triangle allows the chemist the opportunity to distinguish between the two.

In addition, the relatively high dispersant requirement, 5-7.5% (w/w), for formulating liquid and low melting solid technicals as extruded granules may be related to the processing involved: dispersion of the technical among fine particle size silica prior to blending of wetting agent, dispersant, and diluent. The increased surface area that results from the use of silicas may have dictated the additional dispersant requirement in the formulation.

The experimental design blending triangle/tetrahedron was also found to have utility in the identification of the most efficient (lowest application rate) wetting agent and most cost effective diluent for extruded granule. For a carbaryl 75WG, substitution was made at the appropriate triangle apex and the formulation ratios listed above were processed and evaluated. The wetting agent, which performed acceptably at the lowest percent in formulation (w/w), was classified as the most efficient.

Finally, the experimental design triangle/tetrahedron can also be used as a "map" for formulation selection based upon cost vs. performance since the cost of each excipient ratio (per pound of formulated product) can be calculated and superimposed over the area demonstrating acceptable performance.

Conclusions

High concentration extruded granule development using sodium lignosulfonate and sodium naphthalene sulfonate formaldehyde condensate dispersants may be problematic for two primary reasons.

First, the effective dispersant concentration (as dictated by the formulation and dispersant chemical class) must be met in order for the formulation to perform acceptably across a range of water hardnesses and after extended storage at room and elevated temperatures. With sodium lignosulfonate and sodium naphthalene sulfonate formaldehyde condensate dispersants, this concentration range is 3.0-7.5% (w/w) for formulations containing 7.5% or more of wetting agent. For formulations containing 3% wetting agent, the effective dispersant requirement appeared to increase to 9.0% (w/w) for both these dispersant types.

Second, the hydrophobic/hydrophilic surface properties of the formulated technical (and the wetting properties of the associated excipients) will affect the actual wetting agent requirement for a given formulation; the more hydrophobic the surface properties, the greater the requirement. However, wetting agents do not perform with the same efficiency in much the same manner that dispersants do not perform with the same efficiency; the less efficient the wetting agent used, the greater its required concentration in the extruded granule formulation in order to perform effectively. Under the conditions of this research, effective wetting agent requirements for granules made using sodium lignosulfonate and sodium naphthalene sulfonate formaldehyde condensate dispersants ranged between 7.5 - 9.0% (w/w) for formulations containing less than 9% dispersant.

From this research it can be seen that styrene acrylic polymer dispersants can offer significant opportunities in the development of high concentration extruded granules where effective concentrations of sodium lignosulfonate and sodium naphthalene sulfonate formaldehyde condensate dispersants may not "fit" within the available formulation "space". As a practical matter, this is a function of both the desired formulation assay and the assay of the available technical active ingredient. When a total surfactant concentration (wetting agent + dispersant) of 12-15% (w/w) is required for adequate suspension and disintegration of the extruded granule, a target formulation assay of 80% or greater offers very little available space for necessary diluents.

Based on the data developed from this research, a formulation assay of 90% becomes unachievable for naphthalene sulfonate formaldehyde condensate and lignin sulfonate based granule formulations under these conditions. Unlike the other disperant types evaluated in this research, the combined surfactant (styrene acrylic dispersant + wetting agent) requirement for extruded carbaryl technical 75 granules with acceptable performance was 6.0 percent. The general trend toward low overall surfactant requirement appeared to be consistent in the evaluation of other technical active ingredients in extruded granules and is related to a low (3-5%) dispersant requirement for styrene acrylic polymer dispersants under the conditions of evaluation.

Finally, in this era of too much work and not enough time, use of experimental design approaches provides the formulation chemist with an expeditious process for extruded granule development. While in this work several extruded granule formulations were screened that lacked either wetting agent or dispersant or both, the purpose of their inclusion in this fundamental work was to establish that an absolute minimum should be assumed in the absence of actual data. A bona fide effort must be made through experimental design to permit the data to speak for itself and establish the baseline for performance for all components independently before a conclusion can be drawn about their interaction. Having done so at the outset of this (or any) designed experimentation and clearly established the need for all components to be present, such work need not be repeated; the data obtained provides the necessary basis for removing such points from future design consideration. The implication for experimental design is that such work provides a background against which future experimentation can be structured. By design this maximizes learning and minimizes repetition in future related work.

Timothy H. Anderson[1] and Hans-Georg Mainx[2]

Alkoxylated Glyceride Emulsifiers in Agricultural Applications

Reference: Anderson, T. H., and Mainx, H. G., **"Alkoxylated Glyceride Emulsifiers in Agricultural Applications,"** *Pesticide Formulations and Application Systems: A New Century for Agricultural Formulations, Twenty First Volume, ASTM STP 1414*, J. C. Mueninghoff, A. K. Viets, and R. A. Downer, Eds., American Society for Testing and Materials, West Conshohocken, PA, 2001.

Abstract: Ethoxylated glycerides have long been known as effective and safe nonionic emulsifiers. They have been used in applications as diverse as food processing, textile applications, fiber finishes, leather processing, metal working, cosmetics, pharmaceuticals, and of course, as agricultural *other ingredients*. In addition to their excellent emulsifying properties for oils, fats and solvents, ethoxylated triglycerides are widely recognized for their favorable human and ecotoxicological profiles. They are a preferred choice whenever sensitive environments must be protected and in cosmetic applications where they are known to mitigate the irritation potential of more aggressive surfactants.

Current United States Environmental Protection Agency regulations restrict the choices available to agricultural formulators to a single triglyceride, castor oil, with between 5 and 54 moles of ethoxylation on growing crops, raw agricultural commodities and animals. No other ethoxylated or alkoxylated glycerides are specifically approved or located on the "List of 2500", although some companies may have letters in their files allowing specific usage. Unfortunately the agricultural formulator choosing a castor-based surfactant is exposed to cyclical price swings. This would be lessened greatly if alternatives based on soybean or canola oil were found to be satisfactory.

The present work examines the emulsification properties of several alkoxylated glycerides derived from canola oil. Comparisons are drawn with examples of castor oil and castor wax ethoxylates. The systems studied included a selection of solvent-based Emulsifiable Concentrate (EC) formulations as well as a commonly used methyl ester.

Keywords: Pesticide formulations, castor oil ethoxylate, rapeseed oil ethoxylate

Introduction

In the science of agricultural formulations, the current regulatory environment is

[1] Senior Development Chemist, Agrosolutions Group, Cognis Corporation, 4900 Este Ave. Cincinnati, OH 45232.

[2] Doctor, Head of Technical Sales, Cognis Deutschland GmbH, Henkelstrasse 67 D-40551 Dusseldorf, Germany.

dictating much of the strategy for finding solutions. As many longstanding and reliable chemistries have become regulatory targets, the search for alternatives becomes more critical. Castor oil ethoxylates have been called upon (as well as others) to fill these technological gaps. Although success has been achieved, the fact that castor oil is not derived from a commodity source has made pricing uncertain. The experiments detailed in this work were designed to test the interchangeability of seed oil (triglyceride) ethoxylates. Emulsifiable concentrates containing emulsifier blends, in which castor oil ethoxylates play a role, have been formulated using similar ethoxylates based on canola (or rapeseed) oil. The emulsion stability and quality were compared with standard samples.

Castor Oil		Canola Oil	
Ricinoleic Acid	(89.5%)	Oleic Acid	(58%)
Linoleic Acid	(4.2%)	Linoleic Acid	(26%)
Oleic Acid	(3.0%)	Linolenic Acid	(10%)
Palmitic Acid	(1.0%)	Stearic Acid	(6%)
Stearic Acid	(1.0%)		
Other Acids	(1.3%)		

Figure 1-*Typical compositions.*

The typical composition of both castor and canola (rapeseed) oils is outlined above (Figure 1). The most notable feature is the 12-hydroxy present in castor oil (Figure 2). It is the only seed oil containing this monoglyceride (Ricinoleic Ester) and at 89.5% contributes substantially to the unique qualities castor-based surfactants possess. Canola oils typically contain a majority of oleic and linoleic acids, 58% and 26%, respectively (Figure 3). The differences in structure have a direct impact on molecular weight and therefore the calculated HLB (see Table 1).

Recently published studies point out that the 12-hydroxy is not the primary site of ethoxylation for castor-based surfactants (O'Lenick and Parkinson 1993). Cox and Weerasooriya have demonstrated that with proprietary catalysts present one can achieve an insertion reaction with very little secondary ethoxylation (Cox and Weerasooriya 2000). However, under the most commonly used synthetic paths, using alkali or alkali

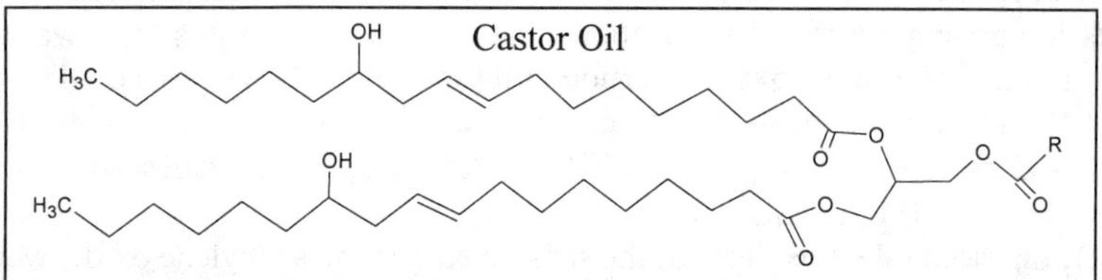

R= 68.5% Ricinoleic, 12.6% Linoleic, 9.0% Oleic, 3.0% Palmitic, 3.0% Stearic, 3.9% Others

Figure 2- *Castor oil fatty acids.*

earth metal catalysts, the presence of secondary ethoxylates or rearrangements (at the 12-hydroxy position) cannot be entirely overlooked. Other seed oils, which lack ricinoleic esters, cannot form this complex mixture of ethoxylates. This disparity in structure will lead to inevitable emulsion performance deviations, but as the greater portion of these molecules is identical, one can expect some similarities to exist.

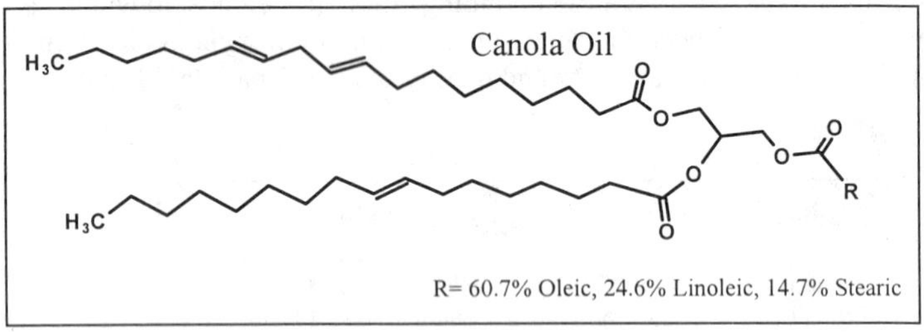

Figure 3- *Canola oil fatty acids.*

Table 1- *HLB comparison*

Moles EO	5	10	20	30	40	60
HLB[1] Castor Oil	3.8	6.4	9.7	11.7	13.1	14.8
HLB[2] Canola Oil	4.0	6.7	10.0	12.0	13.3	15.0

1 Calculated using molecular weight 928.5
2 Calculated using molecular weight 882.1

$$\text{HLB} = \frac{44\text{g}/_{\text{mol EO}}(\text{mol EO})(20)}{\text{molwt of oil} + 44\text{g}/_{\text{mol EO}}(\text{mol EO})}$$

Experimental

Five different agricultural formulations were chosen representing insecticides and herbicides. The pesticides were chosen because the products had emulsification systems that contained castor oil ethoxylates. Each of the emulsifier blends had between 2 and 10 components comprising the complete functional blend. The scope of this work was to substitute only the castor oil ethoxylate portion of these blends, with rapeseed oil ethoxylates. All other components in each blend were held constant. The castor oil ethoxylates used in these experiments were TRYLOX® brand and the rapeseed oil ethoxylates were AGRIMUL RSO® brand.

In phase1, rapeseed oil ethoxylates of the same average moles ethylene oxide, were formulated as a substitute for castor oil ethoxylates. The emulsification performance was evaluated and reported. As necessary, formulations were explored over a range of ethoxylation to optimize emulsion performance (Phase 2).

Methods

Emulsion stability was evaluated using visual inspection of diluted samples as well as measurement of emulsion droplet size. All of the emulsion droplet size data was determined using a Malvern Mastersizer S, laser diffraction instrument. This instrument has an effective range of 0.05µm to 900 µm. The data is reported as the Sauter Mean, D(3, 2), that the authors believe gives data more representative of the entire droplet size distribution. The emulsion stability was determined by one of two methods outlined below. The methods are variations of CIPAC and ASTM methods, modifications being made to allow the most accurate data collection.

All samples were diluted as 5.0g formulation in 95.0g prepared water. The water was prepared as 20 ppm, 342 ppm and 500 ppm hardness (calcium/magnesium) according to the WHO method. The first deviation from published method was that all tests were run with water and formulation at laboratory temperature (23.1C). As a standard formulation was simultaneously tested (for direct comparison), it was felt that accurate conclusions could be drawn.

Preliminary testing indicated the formulations that produced top cream/oil or bottom cream/sediment. Based on the preliminary results emulsion tubes were chosen to maximize the ability to detect layering at all observation times. Conical bottom (centrifuge type tubes), with 0.05 mL increments, were used for bottom layering formulations and the standard straight stoppered tubes were used for top layering formulations.

The observation times were chosen to be 15 min, 30 min, 60 min, 2 hours and 24 hours. All observations were made with a 40-watt incandescent inspection lamp and reported as the nearest tenth of the smallest marked division.

Each emulsion was qualitatively assessed at the beginning of each test (bloom and emulsion quality) and at the conclusion. Bloom was evaluated as spontaneous emulsification based on the ASTM Method E 1116-86 (gravity only). Sample tubes were inverted ten times and then emulsion quality was judged. At the conclusion of 24 hours, sample tubes were inverted until the emulsion was homogenized. The number of inversions is reported as well as another evaluation of the emulsion quality.

The emulsion droplet size was measured after all other stability testing was completed. The numbers reported reflect the 24-hour droplet size after minimal (1-15 inversions) agitation.

Formulations

Table 2 summarizes the different formulations prepared, during phase 1, for emulsion stability testing.

Table 2- *Pesticide formulas phase 1*

Ingredient		% wt/wt	Ingredient		% wt/wt
Active	Chlorpyrifos	46.0	Active	Chlorpyrifos	46.0
Triglyceride	Castor oil POE-45	0.36	Triglyceride	RSO[1] POE-45	0.36
Total emulsifier		7.50	Total emulsifier		7.50
Active	Trifluralin	48.03	Active	Trifluralin	48.03
Triglyceride	Castor oil POE-45	0.22	Triglyceride	RSO POE-45	0.22
Total emulsifier		7.60	Total emulsifier		7.60
Active	2,4-D IOE	95.20	Active	2,4-D IOE	95.20
Triglyceride	Castor oil POE-36	1.32	Triglyceride	RSO POE-36	1.32
Total emulsifier		4.80	Total emulsifier		4.80
Active	Methyl Oleate	85.00	Active	Methyl Oleate	85.00
Triglyceride	Castor oil POE-25	3.75	Triglyceride	RSO POE-36	3.75
Total emulsifier		15.00	Total emulsifier		15.00
Active	Permethrin	10.00	Active	Permethrin	10.00
Triglyceride	Castor oil POE-16	15.02	Triglyceride	RSO POE-16	15.02
Total emulsifier		26.50	Total emulsifier		26.50

[1] RSO stands for rapeseed oil.

Table 3 summarizes the different formulations prepared, in phase 2, after two of the initial emulsions failed. The 2,4-D isooctyl ester was the most difficult formulation to emulsify with Rapeseed oil ethoxylates. All of the RSO formulations between POE-20 and POE-40 were screened, with little to no change. Table 3 represents the best formulation of those efforts.

Table 3- *Pesticide formulas phase 2*

Ingredient		% wt/wt	Ingredient		% wt/wt
Active	2,4-D IOE	95.20	Active	Methyl Oleate	85.00
Triglyceride	RSO[1] POE-30	1.32	Triglyceride	RSO POE-23	3.75
Total emulsifier		4.80	Total emulsifier		15.00

[1] RSO stands for rapeseed oil

Results and Discussion

Nine of the recipes from Phase 1 yielded formulations with identical physical appearance, whether formulated with castor oil ethoxylates or rapeseed oil ethoxylates. The biggest variation was seen when the Permethrin was formulated with rapeseed oil POE-16. This formula is a water-in-oil emulsion, which normally appears clear, amber. The standard castor oil POE-16 formula appeared normal, but the RSO POE-16 product was significantly more viscous and milky white in appearance. Several attempts to reformulate the product (changing order of addition and using a range of RSO POE) did not change its physical appearance. Subsequent dilution tests confirmed that the product still had oil as the continuous phase, however the rapeseed oil ethoxylates were found incapable of formulating a clear, homogenous product.

Tables 4 to10 summarize the emulsion stability results for the formulated samples. It will be noted that the Permethrin sample of abnormal appearance was also included in the testing. This discussion of results will focus only on the comparative performance of the formulations cited in Tables 4 to10. No attempt will be made to correlate this data with any product standards currently available in the industry.

Chlorpyrifos

Apart from deviation in the total cream/sediment after 24 hours, these two formulations performed equally in all waters. The rapeseed oil ethoxylate showed higher levels after 24 hours. No Phase 2 evaluations were made of these formulations.

Table 4- *Chlorpyrifos EC emulsion performance*

Conical bottom Tube	Chlorpyrifos CO POE-45			Chlorpyrifos RSO POE-45		
	20 ppm	**342 ppm**	**500 ppm**	**20 ppm**	**342 ppm**	**500 ppm**
Bloom	Excellent	Excellent	Excellent	Excellent	Excellent	Excellent
Emulsion	Excellent	Excellent	Excellent	Excellent	Excellent	very good
15 min	0.20 mL	Trace sed.	0	0.20 mL	0	0
30 min	0.7 mL	0.20 mL	0.15 mL	0.7 mL	0.20 mL	0.10 mL
60 min	1.5 mL	0.40 mL	0.30 mL	1.5 mL	0.40 mL	0.25 mL
2 hour	2.0 mL	0.70 mL	0.6 mL	2.0 mL	0.7 mL	0.6 mL
24 hour	2.8 mL	1.6 mL	1.9 mL	3.0 mL	1.8 mL	1.8 mL
	No oil	no oil	no oil	no oil	no oil	no oil
Reinversions	3	2	2	4	2	2
Emulsion	Good	Excellent	Excellent	Fair	Excellent	Excellent
Sauter Mean		3.88 μm			8.94 μm	

Trifluralin

Apart from deviation in the total cream/sediment after 24 hours, these two formulations performed equally in all waters. The castor oil ethoxylate showed higher levels after 24 hours. In fact during the entire test the rapeseed oil ethoxylate formulation showed slightly less layering. However due to the very fine divisions in the sedimentation tubes, we are not inclined to place significance on these variations. No Phase 2 evaluations were made of these formulations.

Table 5- *Trifluralin EC emulsion performance*

Conical bottom Tube	Trifluralin CO POE-45			Trifluralin RSO POE-45		
	20 ppm	**342 ppm**	**500 ppm**	**20 ppm**	**342 ppm**	**500 ppm**
Bloom	Excellent	Excellent	Excellent	Excellent	excellent	Excellent
Emulsion	Excellent	Excellent	Excellent	Excellent	excellent	Excellent
15 min	Trace sed	0	0	Trace sed	0	0
30 min	0.05 mL	Trace sed	trace sed	Trace sed	trace sed	trace sed
60 min	0.15 mL	Trace sed	trace sed	0.05 mL	trace sed	trace sed
2 hour	0.35 mL	0.05 mL	0.6 mL	0.15 mL	trace sed	trace sed
24 hour	1.4 mL	0.5 mL	0.4 mL	1.0 mL	0.35 mL	0.25 mL
	No oil	no oil	no oil	no oil	no oil	no oil
Reinversions	4	3	3	4	4	4
Emulsion	Excellent	Excellent	Excellent	Excellent	excellent	Excellent
Sauter Mean		1.16 μm			1.13 μm	

Methyl Oleate

It will be immediately evident that straight substitution of the rapeseed oil ethoxylate did not perform equally to the castor oil ethoxylate. Product formulated in Phase 2 showed only slight improvement even though a range of ethoxylates from 20 POE to 40 POE was investigated. The RSO POE-23 reported showed the closest comparison to the castor oil standard. Over all this particular formulation shows no prospect of direct substitution.

Table 6- *Methyl Oleate emulsion performance phase 1*

Straight Tube	Methyl Oleate CO POE-25			Methyl Oleate RSO 25		
	20 ppm	**342 ppm**	**500 ppm**	**20 ppm**	**342 ppm**	**500 ppm**
Bloom	Fair	Fair	Fair	Good	Good	Good
Emulsion	Good	Good	Very good	Very good	very good	Good
15 min	0	0	0	1.0 mL	0.5 mL	0
30 min	0	0	0	1.5 mL	0.75 mL	Trace sed
60 min	0	Trace sed	0	2.0 mL	1.0 mL	0.5 mL
2 hour	0.5 mL	0.5 mL	Trace sed	2.5 mL	1.5 mL	1.5 mL
24 hour	5.5 mL	2.5 mL	2.2 mL	4.0 mL	4.0 mL	3.5 mL
	Free oil	Free oil	free oil	Free oil	free oil	free oil
Reinversions	1	1	1	1	1	1
Emulsion	Very good	Very good	Very good	Very good	very good	Very good
Sauter Mean		7.85 μm			11.82 μm	

Table 7- *Best Methyl Oleate emulsion performance phase 2*

Straight Tube	Methyl Oleate RSO 23		
	20 ppm	**342 ppm**	**500 ppm**
Bloom	Good	Good	Good
Emulsion	very good	very good	very good
15 min	0.5 mL	trace sed	0
30 min	1.0 mL	0.5 mL	0
60 min	1.5 mL	1.0 mL	trace sed
2 hour	2.0 mL	1.5 mL	0.5 mL
24 hour	2.5 mL	4.0 mL	2.5 mL
	free oil	Free oil	Free oil
Reinversions	1	1	1
Emulsion	Very good	Very good	Very good
Sauter Mean		12.60 μm	

2,4-D Isooctyl Ester

These formulations showed a poor comparison. Phase 2 formulations, ranging from RSO POE-20 to RSO POE-40 did little to improve the situation. The only slight improvement, with RSO POE-30 is reported. The emulsion droplet (Malvern) data shows significant differences at the highest end of the size distribution. Even when re-dispersed these products had poor emulsion stability.

Table 8- *2,4-D Isooctyl Ester emulsion performance phase 1*

Conical bottom Tube	2,4-D IOE CO POE-36			2,4-D IOE RSO POE-36		
	20 ppm	342 ppm	500 ppm	20 ppm	342 ppm	500 ppm
Bloom	Fair	Good	Good	Poor	Good	Good
Emulsion	Very good	very good	Very good	Poor	Fair	Fair
15 min	2 mL	0.1 mL	0.4 mL	4.5 mL	4.0 mL	4.0 mL
30 min	2.3 mL	0.2 mL	0.7 mL	5.0 mL	4.5 mL	4.5 mL
60 min	2.6 mL	0.35 mL	1.2 mL	6.0 mL	5.0 mL	5.0 mL
2 hour	3.0 mL	0.6 mL	1.5 mL	6.8 mL	5.9 mL	5.0 mL
24 hour	4.3 mL	1.8 mL	3.0 mL	11.0 mL	6.5 mL	5.5 mL
	No oil	no oil	No oil	no oil	no oil	No oil
Reinversions	10	9	8	10	10	10
Emulsion	Good	very good	Very good	Good	Fair	Fair
Sauter Mean		6.15 um			126.97 um	

Table 9- *Best 2,4-D Isooctyl Ester EC emulsion performance phase 2*

Conical bottom Tube	2,4-D IOE RSO POE-30		
	20 ppm	342 ppm	500 ppm
Bloom	Poor	Good	fair
Emulsion	Fair	Poor	poor
15 min	4.0 mL	2.8 mL	3.0 mL
30 min	4.5 mL	3.2 mL	3.8 mL
60 min	5.0 mL	4.0 mL	4.5 mL
2 hour	6.0 mL	4.5 mL	5.0 mL
24 hour	10.5 mL	6.5 mL	6.5 mL
	No oil	no oil	no oil
Reinversions	10	15	15
Emulsion	Good	Poor	poor
Sauter Mean		84.23 um	

Permethrin

This formulation was added to the test in order to start to investigate other formulation types. This is a water-in-oil emulsion, which is diluted for application. It is formulated with castor oil ethoxylates, much like a micro-emulsion system. The results previously mentioned were a very different physical appearance both as a concentrate and upon dilution. The emulsion droplet size data also shows a ten-fold reduction in diluted droplet size for the castor oil product. All of the factors for this drastic change are not fully understood at this time. This research will demand much more attention than has been given here.

The one factor that can be seen as a possible correlation is the percentage of alkoxylated triglyceride in the emulsifier blend and in the total pesticide formulation. The formulations with the best performance all had less than five-percent alkoxylated triglyceride in the total surfactant blend.

Table 10- *Permethrin W/O emulsion performance phase 2*

Straight Tube	Permethrin CO POE-16	Permethrin RSO POE-16
	500 ppm	500 ppm
Bloom	Poor	Good
Emulsion	Clear	Fair
15 min	0	0
30 min	0	0
60 min	0	0
2 hour	0	0
	No oil	No oil
Reinversions	0	0
Emulsion	Clear	Fair
Sauter Mean	0.27 um	1.53 um

Conclusions

Based on our limited research, rapeseed oil ethoxylates have a limited future as replacements for castor oil ethoxylates. The interchangeability has limits, as have been discussed, and much more work has yet to be done. The formulations that presented the greatest challenge all had alkoxylated triglyceride formulated as at least 25% of the total surfactant. In systems where the alkoxylated triglyceride is called upon to affect a greater amount of the emulsification a more significant impact to performance should be expected.

If the emulsifier blend contains greater than 20 weight percent of castor oil ethoxylate, there will be a loss in performance when substituted with a comparable rapeseed oil ethoxylate. For emulsifier blends with castor oil ethoxylate between 10 weight percent and 20 weight percent successful substitution with a rapeseed oil ethoxylate will be active dependent. However as many systems have emulsifier blends with weight percentages of castor oil ethoxylates below ten, it is these systems where equal or better performance can be expected when rapeseed oil ethoxylates are substituted.

References

Cox, M. F., Weerasooriya, U., April 2000, "Partially Saponified Triglyceride Ethoxylates," *Journal of Surfactants and Detergents*, Vol. 3, No. 2, pp. 213-220.

O'Lenick, A. J., Parkinson, J. K., November/December 1993, "Group Selectivity of Ethoxylation of Hydroxy Acids," *Journal of the Society of Cosmetic Chemists*, Vol. 44, pp. 319-328.

SESSION V: Regulatory Issues

Ellen M. Mihaich,[1] Carter G. Naylor,[2] and Charles A. Staples[3]

Environmental Management Strategies for Biodegradable Nonylphenol Ethoxylates in Agricultural Products

---

Reference: Mihaich, E. M., Naylor, C. G., and Staples, C. A., "**Environmental Management Strategies for Biodegradable Nonylphenol Ethoxylates in Agricultural Products,**" *Pesticide Formulations and Application Systems: A New Century for Agricultural Formulations, Twenty First Volume, ASTM STP 1414*, J. C. Mueninghoff, A. K. Viets, and R. A. Downer, Eds., American Society for Testing and Materials, West Conshohocken, PA, 2001.

Abstract: Nonylphenol ethoxylates (NPEs) are surfactants used as inert ingredients in pesticides and spray adjuvants. NPEs and their biodegradation intermediates [low mole ethoxylates and ether carboxylates (NPEC)] enter the terrestrial and, to a lesser extent, the aquatic environments through these uses. Nonylphenol (NP) is the basic building block of NPEs and can be a minor biodegradate. Responding to proposed water quality limits in the United States and Canada, the Alkylphenols & Ethoxylates Research Council initiated the NPE Environmental Management Program (EMP). The program, discussed in this paper, includes the development of voluntary, risk-based water quality levels for NPE, NP, and NPEC, and monitoring of waterways to determine whether water quality guidelines are being met; it also provides technical expertise to determine appropriate control methods. The approaches used in the EMP are applicable for most organic compounds and should be an integral component of best management practices.

Keywords: Water quality, inerts, formulation, risk assessment

---

[1]Senior Environmental Toxicologist, Rhodia Inc., 5171 Glenwood Ave., Suite 402, Raleigh, NC, 27612.
[2]Senior Research Associate, Huntsman Corporation, PO Box 15730, Austin, TX, 78761.
[3]Principal Environmental Scientist, Assessment Technologies, Inc, 10201 Lee Highway, Suite 305, Fairfax, VA, 22030.

Introduction

Alkylphenol ethoxylates (APE) are commodity surfactants used in many cleaning products and industrial processes, such as the production of pulp and paper, textiles, leather, paints and coatings, and metal finishing. In the agricultural products industry, alkylphenol ethoxylates, primarily nonylphenol ethoxylate (NPE), are used for their surfactant or surface-active properties. They are used as emulsifiers for pesticides and animal feeds, spreaders, and irrigation aids. In formulated pesticides, NPE surfactants help the formulated pesticide product to disperse and to facilitate the delivery of the active pesticide ingredient to the intended target. By using surfactants, the pesticide product is evenly distributed, thus allowing for a more targeted delivery and a reduction in the amount of the active ingredient needed for effective control. NPE surfactants are mostly classified as "list 4B inerts" (60 FR 35396) under the Federal Insecticide, Rodenticide and Fungicide Act (FIFRA), which means that they may be widely used on both food and non-food crops. They are inexpensive and are extremely effective products that have been used for over 50 years. The knowledge base of these products' environmental and human health properties is large and more extensive than exists for other, newer surfactants that serve similar purposes.

A mounting number of initiatives around the world addressing alkylphenol and alkylphenol ethoxylates [most notably NPE and nonylphenol (NP), the basic building block of NPEs and a minor degradation product] have focused attention on their use and disposal. The main areas on which scientific assessments of NPE/NP focus are generally the aquatic and terrestrial environments. Most recently, the specter of alkylphenols as potential disrupters of the endocrine systems of man and animals is being raised with increasing frequency in the press and in governmental forums, and has spawned considerable scientific research. In response to the growing regulatory and public interest in these products and to ensure proper product stewardship, the APE Research Council (APERC), Washington, DC, initiated the NPE Environmental Management Program (EMP). Its purpose is to assure formulators, users, and regulators of NP and NPEs that these compounds can be safely and effectively used. Not only does this program benefit the users of APEs, the implementation of the best management practices in the EMP will effectively reduce emissions of any organic compound into the environment.

The agricultural products industry is one of the sector industries for which the APE Research Council is developing guidelines under the EMP. There are two distinct types of emissions to the environment of organic compounds relating to agricultural products. The first is from facilities that manufacture or formulate products. Release from manufacturing or processing facilities may occur via permitted effluents following onsite treatment or via treatment in municipal treatment plants. The second way compounds enter the environment is from the intended use of the agricultural product in the environment. Use of the EMP assures the manufacturer and processors, as well as regulators and the public of the environmental safety of the use of these compounds. The objectives then of this paper are to: 1) introduce the EMP and its key components, 2) briefly review critical data on toxicity, wastewater treatability, environmental fate, and monitoring data that support the EMP, and 3) show how the EMP can assist in the management of emissions of NPE surfactants to the environment from their manufacture, processing, and use.

Environmental Management Program

The Alkylphenols & Ethoxylates Research Council developed the NPE Environmental Management Program in early 2000 to help companies meet forthcoming water quality guidelines on NP and its ethoxylates in the United States and potential future regulations in Canada. The Program is designed to support the continued use of NPE-based products by promoting product stewardship and wastewater treatment practices. This program is structured around the following four components:

1) Development of voluntary, risk-based water quality levels for NPE, NP and related transitory degradation intermediates of NPE to complement guidelines being developed by regulatory agencies;
2) Development of practical means for formulators and users to comply with the proposed and/or voluntary water quality guidelines or levels by using widely accepted product stewardship and wastewater treatment practices;
3) Monitoring of targeted waterways in conjunction with government agencies to help determine whether water quality guidelines are being met; and,
4) Providing technical expertise to manufacturers, formulators and end-users to determine appropriate analytical methods and control methods.

Water Quality Guidelines

In the United States, the Environmental Protection Agency (EPA) has proposed a draft national ambient water quality criteria for nonylphenol of 5.9 μg/L (U.S. EPA 2000) for the long term protection of freshwater aquatic life, referred to as the Criterion Continuous Concentration (CCC). The draft criteria for protection against short-term exposures is about 22 μg/L for NP, referred to as the Criterion Maximum Concentration (CMC). The abundant database of acute aquatic studies on NP was used to develop these criteria (Staples et al. 1998, Servos 1999). First, the Final Acute Value or FAV was calculated that reflects the lowest 95th%ile value of the database. If fewer than 20 acute studies are available, then the FAV is the lowest value. The CMC is calculated as the FAV/2. Acute to chronic ratios (ACRs) were calculated from acute and chronic studies conducted on the same species. The FAV divided by the Final ACR (geometric mean of all ACRs) is the Final Chronic Value (FCV). The CCC equals the FCV. For NP, the FAV was calculated to be 44 μg/L and the CMC is 22 μg/L. The draft final ACR was 7.6 so the FCV was calculated to be 44/7.6 or 5.9 μg/L for NP and becomes the CCC. The draft criteria are still under internal review at EPA, but will be used to illustrate how they will aid the EMP.

In Canada, the recent risk assessment for NPE and NP was prepared according to guidance for their Priority Substances List 2 (PSL2) substances under their Canadian Environmental Protection Act (CEPA). As part of this assessment, Environment Canada developed criteria for the long-term protection of aquatic life for NP and for commercial NPE and the most common degradation intermediates, NPE1,2 and the low mole ether carboxylates, NPEC1-4 (Servos et al. 2000). Using a slightly different approach from the United States, but again relying on the abundant database for NP, the acute studies were all ranked from low to high, an acute to chronic ratio was calculated using all available chronic studies, and a distribution of chronic values (each acute value divided by the average ACR) was developed. The resulting value equal to the lowest 95th%ile was about 10 μg/L. For reasons that are unclear, the value of 10 μg/L was divided by a conservative "application factor" of 10 to become the Expected No Effect Level (ENEC) of 1 μg/L for NP. Environment Canada then reviewed the available data for NPE1,2, for NPE3 and higher ethoxylates, and for NPEC. Using the concept of toxic equivalency, they concluded that proportionally, NPE1,2 were 2-fold less toxic than NP, NPE3 and above were 200-fold less toxic than NP and NPEC1-4 were also 200-fold less toxic than NP. The resulting ENECs protective of long-term exposure of these compounds to aquatic life are shown in Table 1.

Voluntary Guidelines to Protect Aquatic and Terrestrial Organisms

Since NPE is easily biodegraded into mainly low mole ethoxylates and ether carboxylates, as well as NP (see discussion below), APERC believes that it is prudent to develop for these degradation products guidelines similar to those developed for NP in the United States Similarly, aqueous exposure to NPE and their degradation intermediates occurs not just to water column organisms, but to sediment dwellers as well. In addition, NPE and their degradation intermediates are applied to land in sludge amendments, thereby exposing soil organisms. Starting with the draft criteria for long term protection of aquatic life exposed to NP by the EPA, guidelines protective for other degradation intermediates in water, and for sediment and soil dwellers can be calculated.

The toxic equivalency process of Environment Canada (Servos et al. 2000) was used to calculate voluntary aquatic life guidelines for NPE and NPEC. Starting with the EPA criteria of 5.9 μg/L for NP, similar values for NPE1,2, NPE3 and higher, and NPEC1-4, were calculated (Table 1) for application in the United States Voluntary guidelines for NPE, NP, and NPEC that protect sediment dwelling and soil organisms were calculated using the Equilibrium Partitioning (EqP) theory (DiToro et al. 1991) (Table 1).

Targeted Environmental Monitoring

When necessary, targeted waterways will be monitored by affected facilities in conjunction with government agencies and the APE Research Council to help ensure that water quality guidelines are being met. There may be specific locations in which concern exists, such as effluent-dominated streams, that warrant monitoring. In addition, certain types of industries may wish to broaden their confidence that the environment is not being adversely affected. This may include the pesticide inert manufacturing and processing industries.

Surface Water Concentrations - To date, based on monitoring by the APE Research Council, EPA, Environment Canada, and others, it is known that most rivers and receiving streams in the United States and Canada are already well within "safe limits." Naylor et al. (1992) sampled NP and NPE levels in 30 rivers in the United States. The receiving waters were selected from the nearly 5000 river reaches in the United States that receive industrial wastewater or receive wastewater from one or more municipal sewage treatment plants. The statistically based sampling was designed by EPA scientists to be able to state, with a 95% confidence, that no more than 10% of the universe of U.S. river reaches with such discharges will have higher values than determined in the 30 rivers chosen. NP and NPE were not detected (maximum detection limits (MDLs) ranged from 0.06 to 0.11 μg/L) in more than two-thirds of the rivers. The highest concentrations were 0.64 μg/L for NP, 1.2 μg/L for NPE1,2 and 14.9 μg/L for the combined NPE3-17. These are generally consistent with studies performed in Canada. Environment Canada has conducted extensive monitoring for NP and NPE at 35 sites in the Great Lakes Basin and upper St. Lawrence River (Bennie et al. 1997). NP levels were less than 0.01 μg/L in 30 of 38 samples and less than 0.9 μg/L in the remaining samples. Levels of NPE1,2 ranged from nondetected to 10 μg/L. NPEC concentrations ranged from 0.05 to 2.0 μg/L. This suggests that the vast majority of NP/NPE producers, formulators, and users are operating with adequate environmental controls and should be able to meet the proposed voluntary guidelines. It is known that adequate wastewater treatment plant operation is needed to reduce NPE/NP levels appropriately (Naylor 1996, Ahel et al. 1994a). For those facilities that are not properly operated, releases to the environment may be problematic. Therefore, identification of those locations is needed prior to the implementation of appropriate wastewater treatment practices, thereby reducing discharges to acceptable levels for NPE, NP, and NPEC.

Sediment Concentrations - In United States river sediment samples, NP concentrations averaged a fraction of a μg/g, and ranged from non-detected (MDLs ranged from 0.0023 to 0.0029 μg/g, dw) to 3 μg/g, dw (Naylor et al. 1992). NP concentrations in Canadian sediment ranged from less than 1 μg/g, dw (6 of 9 samples) to a single peak of 72 μg/g, dw (Bennett and Metcalfe 1999, Bennie 1999). The highest water and sediment concentrations in both the Canadian and the U.S. studies were reported at sites with a history of severe pollution or were proximate to the outfalls of facilities serving large industrial areas and cities. Secondary (biological) treatment, preferably activated sludge, is sufficient to degrade many organic compounds and remove them from the system. With secondary treatment, metabolites of the higher mole NPEs predominate in the effluent. Carboxylic acid metabolites (i.e., NP1EC and NP2EC) account for the majority of the degradation products while the lower molecular weight NPE oligomers account for less and NP accounts for a very small fraction, generally less than 5% (Ahel et al. 1994a, Naylor 1995, Field and Reed 1996).

Sludge-Amended Soil Concentrations – Die-away of NPE and NPE degradation intermediates were reported by Bennie (1999) for one experimental sludge-amended site in Canada. While NP was detected initially at a concentration of 2.7 μg/g , it declined to below the detection limit (<0.2 μg/g) within 90 days.

Wastewater Treatment and Natural Degradation Processes

The most common manner of disposal and treatment of NPE is via wastewater treatment plants (WWTP). Many common uses of NPE include household and industrial cleaning products, which by design enter the wastewater stream after use. NPE entering WWTPs in raw sewage is generally in the water phase. The first step in a WWTP is to remove solids from the liquids by gravity settling. Few NPE are removed in this

Table 1 - *Draft voluntary environmental protection guidelines for NPE and their biodegradation intermediates in water, sediment, and soil.*

Compound	Voluntary Guideline
NP_{water}	5.9 µg/L[1]
$NP_{sediment}$	2.4 µg/g[2]
NP_{soil}	12.2 µg/g dw[2]
$NPE1,2_{water}$	11.8 µg/L[3]
$NP1,2_{sediment}$	4.8 µg/g[3]
$NPE1,2_{soil}$	24.4 µg/g dw[3]
$NPE3\text{-}16_{water}$	1180 µg/L[4]
$NPE3\text{-}16_{sediment}$	480 µg/g[4]
$NPE3\text{-}16_{soil}$	2440 µg/g dw[4]
$NPEC_{water}$	1180 µg/L[4]
$NPEC_{sediment}$	480 µg/g[4]
$NPEC_{soil}$	2440 µg/g dw[4]

[1] Based on U.S.EPA methods for calculating ambient water quality criteria (Final Chronic value or FCV).

[2] FCV_{soil} or $ENEC_{soil}$ = FCV_{water} or $ENEC_{water}$ x the average soil partition constant for NP (K_{oc} = 41,300 L/kg) x an organic carbon content (foc) of 5% (0.05) for an agricultural soil following application of biosolids or 1% foc for sediments.

[3] Based on water, sediment of soil criteria for NP times a relative toxicity factor of 2 for NPE1,2.

[4] Based on water, sediment, or soil criteria for NP times a relative toxicity factor of 200 for NPE3-16 and NPEC1,2.

process. After this first step, the primary effluent enters aerated biological treatment, usually activated sludge treatment. Following biological treatment, liquids and solids are separated. Liquids may or may not undergo tertiary treatment prior to release as final effluent to surface water. Solids are generally incinerated or disposed on agricultural fields. Many municipalities are moving away from incineration and relying more on the beneficial use of the sludge solids or biosolids as soil improvement aids (Davis 1999,

Walker 1999, Ramboll 1999). Any constituents that leave the WWTP as a component of final effluent or biosolids may have entered the WWTP with the raw sewage but were either incompletely biodegraded or are themselves degradation intermediates of other compounds that entered the WWTP with raw sewage. NPEs are generally easily treated in wastewater treatment plants with removal efficiencies on the order of 90 to 99%, although heavily loaded and inefficient plants have recorded lower removal rates of 50 to 90% (Naylor 1996, Maguire 1999, Crecenzi et al. 1995, DiCorcia and Samperi 1994, and Ahel et al. 1994a). Since some residual NPE, NP, and NPEC may be detected in permitted effluents to surface waters, or in sludge applied to agricultural soils for beneficial reuse, a review of natural processes that further reduce concentrations of these compounds is warranted.

Degradation in Water

The primary biodegradation of commercial NPE (e.g. NPE-9) was reported by Yoshimura (1986). A river die-away test measuring the primary degradation of NPE-9 and NPE-12 (9 and 12 mole ethoxylation) achieved 97 to 99% removal in 10 days. A study by Serak and Yan (1997) reported 87% loss of the radioactivity of ring-labeled NPE9 in 128 days. Quiroga et al. (1996) studied the biodegradability of NPE-15 in a river die-away test. Primary degradation of 85 to 90% was achieved in the 17 day test and by day 6, nearly all the NPE3-15 had disappeared.

The key biodegradation intermediates of commercial NPE (e.g. NPE-9) include low mole ethoxylates NPE-1,2, the low mole ether carboxylates, NPEC-1,2 and lastly NP. Ready biodegradation tests using the Organization for Economic Cooperation and Development (OECD) protocols have been performed for all of these compounds. Data are summarized in Staples et al. (1999). Using OECD method 301F, NP degraded 62% in 28 days, but did not achieve the 60% level within 10 days of achieving the first 10% removal. Using OECD 301B, the modified Sturm test that measures ultimate biodegradation by quantifying carbon dioxide evolution, NPE1.5, NPEC1.5, and NP were tested. By day 28, NPE1.5 achieved 45% theoretical carbon dioxide formation ($ThCO_2$), NPEC1.5 achieved 58 to 65% $ThCO_2$, and NP achieved 49% $ThCO_2$ formation. These data show that NP, low mole NPE, and low mole NPEC, the degradation intermediates of NPE, are all inherently biodegradable. A simulation test using seawater and low concentrations of 14C-NP was conducted at 11°C for 56 days by Ekelund et al. (1993). After a 28 day lag phase, 50% $ThCO_2$ was achieved in the next 28 days. This study shows that even low aqueous concentrations of NP are easily degraded in surface water following an acclimation phase.

Ahel et al. (1994b) measured the aqueous photo-degradation of octylphenol (OP), NP, and NPE-1 in natural waters. For NP, a photolysis half-life of 10 to 15 hours was observed in surface water. At 20 to 25 cm depth in lake water, the half-life was about 1.5 times slower. Thus, photolysis is an important degradation process for NP, OP, and NPE-1 in shallow surface waters.

Degradation in Sediment

Yoshimura (1986) studied the primary biodegradation of NPE-9 and NPE-12 in aerobic sediments. Primary degradation of 96 to 99% was achieved in 30 days. Ekelund et al. (1993) studied the biodegradation of NP using a system of seawater and sediment with low concentrations of 14C-NP. The study was conducted at 11°C for 56 days under both aerobic and anaerobic conditions. The authors reported the mineralization of NP occurring at 1.2% per day under aerobic conditions and 0.6% per day under anaerobic conditions. Heinis et al. (1999) conducted studies with littoral enclosures, which are closed off portions of a natural lake. NP was applied every other day for about 20 days and mixed in the enclosed water. A variety of ecotoxicological assays were performed. After cessation of NP addition, concentrations in the water and sediment were observed for over one year. An average half-life of about 18 hours was reported for surface water and 66 days for sediments.

Degradation in Soil

The biodegradation of NPE in soil has been examined by several authors. In soil columns under aerobic conditions, Kuchler (1995) reported essentially 100% loss of NP and NPE in 28 days. Marcomini et al. (1989) measured the loss of NP in sludge amended soil. Nearly 80% was degraded in the first three weeks. Degradation slowed after that as the NP became adsorbed to the soil and was no longer bioavailable. In soil / compost systems, Trocme et al. (1988) reported 62 to 89% loss of NP in 40 days. In aerobic soils, Hughes et al. (1994) reported an ultimate biodegradation half-life for NPE-9 of about 18 days. Kirchmann et al. (1991) added NP to soil in concentrations expected from the application of sewage sludge to soil and 100 and 1000 times higher. They measured the effects on soil respiration and NP degradation. NP did not affect soil respiration and degraded by more than 90% within 10 to 20 days. Kujawa et al. (1997) amended sandy soil with two different sewage sludges containing high and low concentrations of NPE and NP and conducted lysimeter studies. NPE was detectable in the 0 to 10 cm layer only until day 18 (low NPE concentration sludge) and day 14 (high NPE concentration sludge). By day 30, no NPE or NP was detectable in either system. A study by Ramboll, Ltd. (1999) in coordination with the Danish EPA conducted composting studies with sewage sludge containing NPE. During the tests, NPE was degraded by 78 to 95%. Maguire (1999) reported the results of studies at a Guelph, Ontario WWTP. Biosolids from this plant were applied at 0, 8 (the maximum allowed application rate), 20 and 40 tonnes per hectare to soil and measured NPE and degradation intermediates. Initial concentrations were 450 mg/kg of NP, 15 mg/kg of OP, no NPE-1, 10 mg/kg of NPE-2, 350 mg/kg of NPE-3-17, and <15 mg/kg APEC. No APE were detectable in the 8 t/ha plot after 121 days (July to October). These data led Maguire (1999) to conclude, "…NP and the lower carboxylates…do not appear to be persistent in soil under aerobic conditions."

Summary of the Fate of NPE and NPE Degradation Intermediates

In his review of the persistence of NPE, Maguire (1999) summarized as follows. "The general conclusion in this section is, as expected from the studies of degradation in STPs [sewage treatment plants], that primary biodegradation of NPEs is faster than ultimate degradation in STPs, that primary biodegradation of NPEs is faster than ultimate degradation of more persistent products such as NP1EO, NPE2EO, NPEC1, NPEC2, and NP. Microbial acclimation to such chemicals is required for optimal degradation efficiencies. Sunlight photo-degradation of such products is also expected to be important. In aquatic ecosystems, it appears that parent NPEs are not persistent, but some degradation products may have moderate persistence, especially under anaerobic conditions." Maguire finally concludes that, "The intermediate and final products of metabolism are more persistent than the parent NPEs, but there is no doubt that such chemicals will also be ultimately biodegraded."

Best Industry Practices

As an example of how the EMP may be implemented for a particular industry, this paper examines the industry that formulates inert surfactants and dispersants into agricultural chemicals such as insecticides, fungicides, and herbicides. NPE products represent approximately 17% of the surfactants used in the agricultural industry (APERC 2000). Some common approaches to waste minimization that can be used during the manufacture and formulation of pesticide products include recycling and waste treatment. These and other practices can be supplemented and optimized to meet the needs of a particular process or facility.

In a survey of pesticide manufacturers and formulators, the majority reported using either partial or full recycling of wastewater from their plants. In particular, the waste streams containing the pesticide active ingredient and "inerts" are usually recycled back in to the process or stored on-site until the next manufacturing campaign. This serves both to prevent release of the active ingredient into the environment and gain economic efficiencies. Effluent that is not recycled is typically treated on-site in at least secondary treatment systems prior to discharge to a sewage treatment works. The chief concern of the formulator is prevention of inappropriate releases of the active ingredient. While inerts commingled with the active ingredient will be treated and recycled as well, improved focus on the inerts will clarify the recycled amounts needed, minimize losses, and reduce potential risks to the environment. In comparison to other industry sectors that use NPE surfactants, the agriculture industry uses relatively small quantities of these materials in their products. Nevertheless, careful handling and use of the material, minimization of product use through vigorous recycling practices, use of onsite treatment and at least secondary, if not tertiary treatment, of releases in offsite sewage treatment plants will minimize releases and hence, environmental concentrations of NPE and their degradation intermediates.

For the actual pesticide application, only 5 to 7% of the formulated product is typically made up of surfactant. In order to minimize the volume for packaging, inventory and transportation in the pesticide industry, the smallest effective quantity of surfactant is used. The 5 to 7% surfactant that is in the formulated product is further diluted in the final spray tank mixture that is applied to the field by the farmer or applicator. The final concentration of APE in the spray tank is approximately 0.1% to 0.2% with adjuvant added, assuming 5% surfactant in the liquid emulsifiable or suspension concentrate formulation. This assumes approximately 5 pints are diluted with 25 gallons of water, resulting in an extremely small concentration of surfactant per acre applied.

Continuing to implement the plan, formulating facilities that may need to verify that their actions are minimizing potential environmental risks from all NPE and degradation intermediates, may wish to work with local authorities on a targeted monitoring program. Monitoring plans should use acceptable analytical methods, such as that of Naylor et al. (1992) and Field and Reed (1996). The methods of Naylor and Field are being submitted to ASTM for adoption as official analytical methods. Following development of monitoring data, comparison with voluntary guidelines can assist in demonstrating to the authorities that the environment is being adequately protected. In the end, the facility assures their communities that the environment is being protected, assures their customers that the materials can be produced, handled, transported, and used safely, and they identify any areas that may need further optimization.

Conclusions

Although nearly all data in North America shows that the environment is not being adversely impacted by the current use patterns of APEs, there is the potential for "hot spots" where either practices are poor, treatment is inadequate, or the natural flows of receiving streams are dominated by effluents from municipal treatment plants and industrial facilities. Implementation of the EMP with each industry sector that uses APEs in a manner that has the potential for impact to the environment is critical to the continued use of these products. Due to the regulatory initiatives underway in North America and elsewhere, vigilance is needed by the producers of the materials and the downstream customers who use and ultimately dispose of the waste product to discover and remedy those areas in need of improvement prior to regulatory enforcement. The development and application of EMP programs targeted to specific industry sectors will aid in the prevention of pollution and improvement of the environment.

References

Ahel, M., Giger, W., and Koch, M., 1994a, "Behavior of Alkylphenol Polyethoxylate Surfactants in the Aquatic Environment - I. Occurrence and Transformation in Sewage Treatment," *Water Research,* Vol. 28, No. 5, pp. 1131-1142.

Ahel, M, Scully, F.E. Jr., Hoigne, J, and Giger, W., 1994b, "Photochemical Degradation of Nonylphenol and Nonylphenol Polyethoxylates in Natural Waters," *Chemosphere,* Vol. 28, pp. 1361-1368.

APERC, 2000, Internal Report on Industry Surfactant Usage, Alkylphenol and Ethoxylates Research Council, Washington, DC.

Bennett, E.R. and Metcalfe, C.D., 1999, "Distribution of Alkylphenol Compounds in Great Lakes Sediments, United States and Canada," *Environmental Toxicology and Chemistry*, Vol. 17, pp. 1230-1235.

Bennie, D.T., 1999, "Review of the Environmental Occurrence of Alkylphenols and Alkylphenol Ethoxylates," *Water Quality Research Journal of Canada,* Vol. 34, pp. 79-122.

Bennie, D.T., Sullivan, C.A., Lee, H-B, Peart T.E., and Maguire, R.J., 1997, "Occurrence of Alkylphenols and Alkylphenol Mono-and Diethoxylates in Natural Waters of the Laurentian Great Lakes Basin and the Upper St. Lawrence River," *Science of the Total Environment,* Vol. 193, pp. 263-275.

Crescenzi, C., DiCorcia, A., and Samperi, R., 1995, "Determination of Nonionic Polyethoxylate Surfactants in Environmental Waters by Liquid Chromatography / Electrospray Mass Spectrometry," *Analytical Chemistry*, Vol. 67, pp. 1797-1804.

Davis, B. 1999, "The Effects of Sludge Application on Land and Impact of Organic Chemicals on the Terrestrial Ecosystem: UK Experience," Presented at the STP Workshop in Coordination with the Danish EPA, April/May 1999.

DiCorcia, A. and Samperi, R, 1994, "Monitoring Aromatic Surfactants and their Biodegradation Intermediates in Raw and Treated Sewages by Solid-Phase Extraction and Liquid Chromatography," *Environmental Science and Technology,* Vol. 28, pp. 850-858.

DiToro, D.M., Zarba, C.S., Hansen, D.J., Berry, W.J., Schwartz, R.C., Cowan, C.E., Pavlou, S.P., Allen, H.E., Thomas, N.A., and Paquin, P.R., 1991, "Technical Basis for Establishing Sediment Quality Criteria for Nonionic Organic Chemicals by using Equilibrium Partitioning," *Environmental Toxicology and Chemistry,* Vol. 10, pp. 1541-1587.

Ekelund, R., Granmo, A ., Magnusson, K., Berrgren, M ., and Bergman, A., 1993, "Biodegradation of 4-Nonylphenol in Seawater and Sediment," *Environmental Pollution,* Vol. 79, pp. 59-61.

Field, J.A. and Reed, R., 1996, "Nonylphenol Polyethoxy Carboxylate Metabolites of Nonionic Surfactants in U.S. Paper Mill Effluents, Municipal Sewage Treatment Plant Effluents, and River Waters," *Environmental Science and Technology,* Vol. 30, pp. 3544-3550.

Heinis, L.J., Knuth, M.L., Liber, K., Corry, T.D., and Stay, F.S., 1999, "Persistence and Distribution of 4-Nonylphenol Following Repeated Application in Littoral Enclosures," *Environmental Toxicology and Chemistry,* Vol. 18, pp. 363-375.

Hughes, A.I., Fisher, J., and Brumbaugh, E., 1994, "Biodegradation of NPE in Soil," Proceedings: 4th CESIO World Surfactants Congress, Vol. 4, pp. 365-372, Barcelona, Spain, June 3-7, 1996. Publ. Royal Society of Chemistry, Cambridge, UK, ISBN 0-85404-766-2.

Kirchmann, H., Astrom, H., and Jonsall, G., 1991, "Organic Pollutants in Sewage Sludge," *Swedish Journal of Agricultural Research,* Vol. 21, pp. 107-113.

Kuchler, T., 1995, *Behavior of Surfactants and their Influence on the Mobility of Organic Micropollutants in Sandy Soils with Weak Sorptive Capacities.* Doctoral dissertation. University of Potsdam, Potsdam, Germany.

Kujawa, M., Schnaak, W., and Kuchler, T., 1997, "Occurrence of Organic Pollutants in Sewage Sludge and Influence of Surfactants on their Mobility in Amended Soil," in *Management and Fate of Toxic Organics in Sludge Applied to Land,* Technical University of Denmark, Copenhagen, Denmark. Workshop proceedings, April 30-May 2, 1997.

Maguire, R.J., 1999, "Review of the Persistence of Nonylphenol and Nonylphenol Ethoxylates in Aquatic Environments," *Water Quality Research Journal of Canada,* Vol. 34, pp. 37-78.

Marcomini, A., Capel, P.D, Lichtensteiger, Th., Brunner, P.H. and Giger, W., 1989, "Behavior of Aromatic Surfactants and PCBs in Sludge-Treated Soil and Landfills," *Journal of Environmental Quality,* Vol. 18, pp. 523-528.

Naylor, C.G., 1995, "Environmental Fate and Safety of Nonylphenol Ethoxylates," *Textile Chemists and Colorists*, Vol. 27, No. 4.

Naylor, C.G., 1996, "The Dirt on Detergents," *Industrial Wastewater.* September/October.

Naylor, C.G., Mieure, J.P., Adams, W.J., Weeks, J.A., Castaldi, F.J., Ogle, L.D., and Romano, R.R., 1992, "Alkylphenol Ethoxylates in the Environment," *Journal of the American Oil Chemists Society*, Vol. 69, pp. 695-703.

Quiroga, J.M., Manzano, M.A., Sales, D., and Perales, J.A., 1996, "Biodegradation of Nonylphenol Polyethoxylate (NPEO) in River Water," p. 417-425, Vol. 4. In *Proceedings of the 4th Surfactants Conference*, Barcelona, Spain, June 3-7, 1996. Publ. Royal Society of Chemistry, Cambridge, UK. ISBN 0-85404-766-2.

Ramboll, Ltd, 1999, "Degradation of Xenobiotics by Composting," *Proceedings of LAS Risk Assessment for Sludge-Amended Soils*, RAP9900811. SPT Workshop in Coordination with the Danish EPA 19/20 April, 1999, Copenhagen, Denmark.

Serak, K.R. and Yan, A., 1997, "Aerobic River Die-Away of ^{14}C-Nonylphenol Ethoxylate Standard Method for Biodegradation by a Shake-Flask Die-Away Method," ABC Laboratories, Inc. Columbia, MO, USA. Report No. #42889, 150 pp.

Servos, M.R., 1999, "Review of the Aquatic Toxicity, Estrogenic Responses and Bioaccumulation of Alkylphenols and Alkylphenol Polyethoxylates," *Water Quality Research Journal of Canada*, Vol. 34, pp. 123-177.

Servos, M.R., Maguire, J., Bennie, D. Lee, B., Cureteon, P., Davidson, N. Suttcliffe, R., and Rawn, T., 2000, *Background Document for the PSL-2 Assessment of Nonylphenol and Nonylphenol Polyethoxylates*. Environment Canada.

Staples, C.A., Weeks, J., Hall, J.F., and Naylor, C.G., 1998, "Evaluation of Aquatic Toxicity and Bioaccumulation of C8- and C9- Alkylphenol Ethoxylates," *Environmental Toxicology and Chemistry,* Vol. 17, pp. 2470-2480.

Staples, C.A., Williams, J.B., Blessing, R.L., and Varineau, P.T., 1999, "Measuring the Biodegradability of Nonylphenol Ether Carboxylates, Octylphenol Ether Carboxylates, and Nonylphenol," *Chemosphere,* Vol. 38, pp. 2029-2039.

Trocme, M., Tarradellas, J., and Vedy, J.C., 1988, "Biotoxicity and Persistence of Nonylphenol During Incubation in a Compost-Sandstone Mixture," *Biology and Fertility of Soils,* 5: 299-303.

U.S. Environmental Protection Agency, 2000, "Draft Ambient Aquatic Life Water Quality Criteria: Nonylphenol," Prepared for the Office of Water, Office of Science and Technology, Office of Research and Development; EPA Contract No. 68-C6-0036.

Walker, J., 1999, "Biosolids Management and Risk Assessment in the United States," Presented at the Sewage Treatment Plant Workshop in Coordination with the Danish EPA, April/May 1999.

Yoshimura, K., 1986, "Biodegradation and Fish Toxicity of Nonionic Surfactants," *Journal of the American Oil Chemists Society*, Vol. 63, No.12, pp. 1590-1596.

SESSION VI: Application and Spray Drift

W. Alison Forster,[1] Mark O. Kimberley,[1] and Jerzy A. Zabkiewicz[1]

Pesticide Spray Droplet Adhesion Modeling

Reference: Forster, W. A., Kimberley, M. O., and Zabkiewicz, J. A., **"Pesticide Spray Droplet Adhesion Modeling,"** *Pesticide Formulations and Application Systems: A New Century for Agricultural Formulations, Twenty First Volume, ASTM STP 1414,* J. C. Mueninghoff, A. K. Viets, and R. A. Downer, Eds., American Society for Testing and Materials, West Conshohocken, PA, 2001.

Abstract: Spray droplet adhesion is a consequence of dynamic interactions of formulants within the spray droplet during flight and on impact, physical properties of the droplet (size and velocity), leaf surface morphology and leaf orientation. The objective of this study was to improve on a universal adhesion model incorporating the above factors. The previous spray droplet adhesion model has been extended to include droplet size (ranging from 180 μm to 1000 μm) and droplet velocity (0.63 ms^{-1} to 2.82 ms^{-1} at impact), and made more robust by increasing the number of formulations (dynamic surface tension of droplets at impact range from 20.7 to 72 mN m^{-1}), the types of leaf surfaces (20% acetone contact angles on the leaf surfaces range from 61° to 124°), and the number of droplet impact angles (leaf angles from horizontal to 67.5°). In this study, contact angles using 20% acetone:water accounted for 95% of the differences between the leaf surfaces studied compared to 76% previously observed with water only. The present universal model explains 74% of the deviance. The current model's capabilities and limitations are discussed in relation to the physical parameters measured and observed.

Keywords: pesticide spray, adhesion, modeling

Introduction

The present environmental concerns over agrochemical residues in the atmosphere, water, soil and foodstuffs will not lessen. Improved application and formulation technologies are needed to reduce spray drift, improve retention on-target, enhance uptake and translocation and generally permit lower rates of pesticides to be used. Spray adjuvants provide a recognized means to improve delivery and efficacy of the active ingredient. The problems in predicting formulation behavior, and the cost of empirical field screening, can be reduced by developing a better understanding of the mode of action of adjuvants in spray formulations. This approach in turn lends itself to the development of either empirical or process-driven models that may be incorporated into computer-based decision support systems.

[1]Scientist, Forest Protection, **Forest Research**; Scientist, **Forest Research**; and Scientist and Project Leader, Forest Protection, **Forest Research**, respectively, Private Bag 3020, Rotorua, New Zealand.

The factors involved in spray formulation efficacy on specific plant species are known in principle, but in practice only some specific component parts have been studied in any detail and these have never been amalgamated to form a spray formulation efficacy model (SFEM). A spray formulation efficacy model could take the following form:

SFE = ***fn*** (Deposition, Retention, Uptake, Translocation, Toxicity)

where deposition is the amount of active ingredient (a.i) or spray impacting the target area (i.e., application volume, minus drift), whereas retention is the amount of solution (droplets and satellite droplets) remaining on the plant (i.e., retention is deposition minus run-off and droplet reflection). A review of the factors incorporated in this model and their current stages of development has been presented recently (Zabkiewicz 2000). The focus of the current work is on adhesion, which is a component variable of retention in the above model. Adhesion has been defined as the "stickability" of droplets on *initial* impact (Forster and Zabkiewicz 1994, Stevens et al. 1993). Droplet adhesion is a consequence of dynamic interactions of formulants within the spray droplet during flight and on impact, physical properties of the droplet (size and velocity), leaf surface morphology and leaf orientation. Interest in adjuvant-enhanced formulation efficacy increases the need to be able to quantify effects and interactions between the spray droplet and the plant surface. Understanding adhesion and the interactions between a droplet and a leaf surface should lead to a better definition of desired formulation, and adjuvant, properties. Although leaves are considered as the spray impact surface in this study, a universal adhesion model could have much wider applicability. For example, adhesion of droplets onto wood or insect surfaces could be estimated and enhanced.

Early adhesion research focused on the reflection of water drops on first impact with artificial surfaces (Hartley and Brunskill 1958), modeling a retention factor using advancing and receding contact angles and solution density factors (Furmidge 1962), and dimensional analysis to study droplet adhesion on barley (Lake and Marchant 1983). A model was developed to predict reflected velocity factors for the "first impact" rebound of droplets from a plant surface (Brazee et al. 1991) and similar energy based modeling has been performed more recently (Webb and Holloway 1999, Webb et al. 2000). A model that included terms for velocity, droplet diameter, angle of incidence of the droplet and solution dynamic surface tension (DST), but still excluded leaf character (Stevens et al. 1993), had been the only one to address primary adhesion effects. This model showed that velocity and droplet size gave a high degree of correlation with adhesion and was the starting point for a universal model of adhesion produced by Forster et al. (1998). The initial universal model for adhesion, using leaf angle, water contact angle, and surface tension as independent variables, took the form

$$Adhesion\ (\%\ of\ droplets\ impacting) = 100e^F / (1 + e^F) \qquad (1)$$

and

$$F = a + 2.22(1-\sin[\theta]) - 0.136\,\varphi + 0.187\,(\gamma_w - \gamma_l)^2 \qquad (2)$$

where

a = 8.67 for very large droplets (~ 1000 μm diameter) or 12.75 for medium / large droplets (~ 650 μm diameter), φ = water contact angle (degrees), θ = angle of impact of the droplet with the leaf surface (degrees), γ_w = surface tension of water

(72 mN m^{-1}), and γ_t = surface tension of the solution at the time of impact.

The deviance explained by the model was 65.4%, indicating that there were substantial influences that had not been taken into account. The objective of the present study was to improve the "universal" adhesion model by adding droplet size and droplet velocity at time of impact as variables. The model was also to be made more robust, by increasing the number of formulations, the range of leaf surface types and droplet impact angles studied.

Methods and Materials

Plants

Seedlings (3-year-old) of red maple (*Acer rubrum*), red oak (*Quercus rubra*), and sweetgum (*Liquidambar styraciflua*) were grown outdoors in the ***Forest Research*** nursery. Fully expanded leaves (adaxial and abaxial) from the mid-section of plants were used. Common lambsquarters (*Chenopodium album*) plants were grown from seed in individual pots containing Bloom potting mix (Yates NZ Ltd.) and raised under controlled environment conditions (70% RH and 14 hour photoperiod, ~ 500 μmol/m^2/s; 23°C/15°C day/night) and used at three weeks of age. The adaxial surfaces of common lambsquarters leaves were studied. The leaves investigated represented a range of leaf surface characters (20% acetone contact angles ranged from 62° to 123.8°).

Chemicals and Formulations

The triclopyr (3,5,6-trichloro-2-pyridinyloxyacetic acid) herbicide (Dow Agrosciences) formulations that were used included Garlon® 4 (butoxyethyl ester), Garlon® 3A (triethylamine salt), and triclopyr triethylamine salt plus various combinations of organosilicone (OSi Specialties, Crompton Corporation) and non-organosilicone surfactants (ISP Technologies, Inc.). Herbicide concentrations ranged from 3.2 to 16 g ae/l and surfactants from 0.1 to 2 g/l. The adhesion of water alone was also studied.

Determination of Adhesion

The method for determining adhesion has been fully described previously (Stevens et al. 1992). An impulse-jet droplet generator was used to produce mono-sized drops with average diameters of approximately 169 μm, 360 μm, 543 μm, and 819 μm. Fall distances studied were 200 mm, 400 mm, 800 mm, and 1485 mm, producing droplet velocities ranging from 0.63 ms^{-1} to 2.82 ms^{-1}. Detached leaves (both abaxial and adaxial surfaces for the tree species, adaxial surface of common lambsquarters) were oriented at 0°, 22.5°, 45°, and 67.5° from the horizontal on plastic stages. In reality leaves are attached to stems with petioles and these will have some elasticity. If a large droplet with high momentum impacted a leaf, a proportion of the energy of impact would be absorbed by the leaf bending, which would reduce the likelihood of bouncing. Ten drops were impacted onto each of five replicate leaves, taken from different plants, per formulation, for each droplet size, droplet fall distance, leaf angle, and leaf surface studied. If the droplet stuck, then this was defined as adhesion; if it bounced, regardless of distance, this was defined as no adhesion.

Droplet Velocity

Impact velocity was determined for a droplet from each size class, for every fall distance, by multiple-exposure photography using ultrahigh speed (ISO 20 000) instant film (Polaroid 612) and stroboscopic illumination. The velocity data was modeled, as described in the results and discussion, in order to calculate the velocity of any given droplet size and fall distance.

Contact Angle Measurements

Droplets (2 µl) of 20% (v/v) acetone were applied to adaxial and abaxial surfaces of freshly detached leaves of each of the three tree species and the adaxial surface of common lambsquarters. The contact angle was calculated from the projected image of the droplet. Twenty droplets per formulation for each leaf surface were measured.

Dynamic Surface Tension (DST)

A Krüss Bubble Pressure Tensiometer BP2 MKII was used to measure DST over a surface age time span of 15 to 1000 milliseconds.

Statistical Analysis

Multiple logistic regression was used to model the effect on percentage droplet adhesion of droplet size, leaf angle, leaf surface and formulation. An analysis of deviance was used to measure the effectiveness and statistical significance of each term in the model. Deviance is a measure of variation used in logistic regression models (McCullagh and Nelder 1983) and the analysis of deviance is equivalent to the analysis of variance of a conventional regression model.

Results and Discussion

Droplet Velocity and Surface Age at Time of Impact

Droplet velocity at time of impact with the leaf surface was required for two reasons. First, a velocity factor needed to be added into the adhesion model. Secondly, fall time, and hence droplet age at time of impact, needed be calculated to determine the time at which dynamic surface tension (DST) had to be established.

Two models of droplet velocity, an empirical model and a more theoretically-based model, were tested. The first was a modification of the empirical model used by Stevens *et al.* (1993)

$$V = z - y\, r^{D} \qquad (3)$$

where,

$z = aS^b$, $y = z - c$, $r = (y/z)^d$ if $y > 0$, $r = 1x10^{-20}$ if $y < 0$

V is velocity (ms^{-1}), $\acute{S}$ is droplet diameter (mm), D is fall distance (m), $a = 3.50$, $b = 0.954$, $c = 0.766$, $d = 9.81$.

Fall time as a function of distance can be obtained by integration

$$Fall\ time\ (s) = D/z + (ln(z - y/z - yr^D)/(z\ ln(r)) \quad (4)$$

Although the empirical model for velocity appears to fit the measured data well ($R^2 = 0.97$), a more theoretical approach should be possible. The drag on an object falling through air can generally be assumed to be proportional to the square of velocity. Based on this, an equation for velocity as a function of fall distance was derived by integration. This equation fitted the fall distance versus velocity data well ($R^2 = 0.95$). Drag is also proportional to the cross-sectional area of the falling object. However, when the square of droplet diameter was incorporated into the velocity equation, it did not fit the data well. Instead, it has been assumed that drag is proportional to diameter raised to a power, and the power coefficient has been estimated using nonlinear regression with the velocity data. The following equation is obtained by integration.

$$V = \sqrt{(d/z)}tanh(acosh(exp(zD)cosh(atanh(\sqrt{(z/d)}c))) \quad (5)$$

where

V, D and z are as before; $a = 0.817$, $b = -1.93$, $c = 0.383$, $d = 9.8$

$$Fall\ time = (acosh(exp(zD)cosh(atanh(\sqrt{(z/d)}c))) - atanh(\sqrt{(z/d)}c)) / \sqrt{(zd)} \quad (6)$$

Table 1 shows the velocity predictions from the empirical model (Model 1) and the theoretically based model (Model 2), while Table 2 shows the fall time predictions from each model. The two models gave similar velocity and fall time predictions for the range of droplet sizes and fall distances used in the trial. The second, theoretically based, model was used for calculating velocity and surface tension values in subsequent analyses.

Table 1 — *Velocity (ms^{-1}) predictions from an empirical and a theoretically based model (Models 1 and 2 respectively).*

Height (m)		Diameter (mm)			
		169	360	543	819
Model 1	0.2	0.64	1.22	1.51	1.73
	0.4	0.64	1.30	1.79	2.26
	0.8	0.64	1.32	1.93	2.70
	1.485	0.64	1.32	1.95	2.87
Model 2	0.2	0.62	1.23	1.57	1.79
	0.4	0.62	1.29	1.81	2.26
	0.8	0.62	1.29	1.91	2.64
	1.485	0.62	1.29	1.92	2.82

Table 2 — *Fall time (s) predictions from the empirical and theoretically based models (Models 1 and 2 respectively).*

	Height (m)	Diameter (mm)			
		169	360	543	819
Model 1	0.2	0.31	0.19	0.17	0.16
	0.4	0.62	0.35	0.29	0.26
	0.8	1.24	0.65	0.51	0.42
	1.485	2.31	1.17	0.86	0.66
Model 2	0.2	0.33	0.21	0.19	0.18
	0.4	0.66	0.37	0.30	0.27
	0.8	1.30	0.68	0.52	0.43
	1.485	2.40	1.20	0.87	0.68

Dynamic SurfaceTension

Dynamic surface tension was determined experimentally for each solution and the resulting data modeled so that the surface tension of the droplets at impact could be calculated easily. Figure 1 shows the dynamic surface tensions for the solutions studied.

A three-parameter dynamic surface tension curve was fitted for each formulation so that the surface tension of droplets at the time of impact could be calculated

$$\textit{Surface tension} = S_1 + (72 - S_1)e^{S_2 \cdot t^{S_3}} \qquad (7)$$

where

t is time (ms) and the coefficients used are shown in Table 3.

As shown by the high R^2 values, these equations provided good models of the profiles of dynamic surface tension of the various formulations.

Droplet Surface Tensions at Time of Impact

The dynamic surface tensions at time of impact were obtained by combining the models for fall time and surface tension for each formulation and concentration gave (Table 4). This shows that the surface tension at time of impact was generally greater with increasing droplet size, for any given fall distance. This would be due to shorter fall times, and hence lower surface ages, as larger droplets fall at greater velocities than smaller droplets.

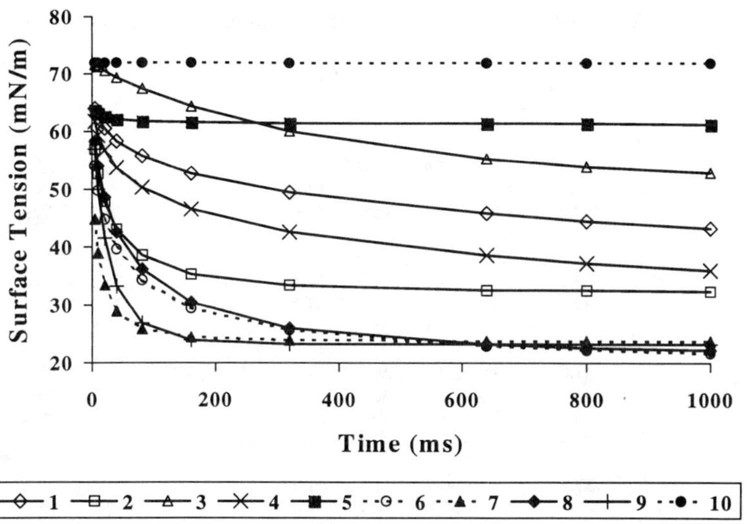

Fig. 1: *Effect of formulation on dynamic surface tension. 1 and 2 are Garlon® 4; 3 and 4 are Garlon® 3A (commercial formulations of triclopyr); 5 is a base triclopyr formulation; 6 and 7 contain triclopyr + organosilicone surfactant + conventional surfactant; 8 and 9 contain triclopyr + organosilicone surfactant. 10 is water alone. Treatments 1, 3, 6 and 8 contain triclopyr at 3.2 g ae / L; treatments 2, 4, 5, 7 and 9 contain 16 g ae/L.*

Table 3 — *Coefficients for, and variance (R^2) accounted for by dynamic surface tension models.*

Product I.D Number[1]	Active Rate	S_1	S_2	S_3	R^2
1	(3.2 g ae/l)	12.7	-0.091	0.288	98.1
2	(16 g ae/l)	32.4	-0.215	0.492	99.3
3	(3.2 g ae/l)	50.4	-0.005	0.881	99.9
4	(16 g ae/l)	25.3	-0.145	0.335	99.7
5	(16 g ae/l)	61.3	-1.009	0.248	99.1
6	(3.2 g ae/l)	20.4	-0.223	0.404	99.7
7	(16 g ae/l)	23.8	-0.387	0.477	99.7
8	(3.2 g ae/l)	21.6	-0.142	0.494	99.7
9	(16 g ae/l)	23.3	-0.119	0.704	98.5

[1] 1and 2 are Garlon® 4; 3 and 4 are Garlon® 3A (commercial formulations of triclopyr); 5 is a base triclopyr formulation; 6 and 7 contain triclopyr + organosilicone surfactant + conventional surfactant; 8 and 9 contain triclopyr + organosilicone surfactant.

Table 4 — *Predicted surface tensions (mN m^{-1}) for formulations at time of impact.*

Fall distance (m)	Diameter (μm)	Formulation 1	2	3	4	5	6	7	8	9
0.2	169	49.1	33.3	60.5	42.3	61.5	25.4	23.9	25.7	23.4
	360	51.4	34.4	63.4	44.9	61.6	27.9	24.2	28.5	23.6
	543	52.0	34.8	64.1	45.6	61.6	28.6	24.3	29.3	23.8
	819	52.2	35.0	64.4	45.9	61.6	28.9	24.3	29.7	23.9
0.4	169	45.5	32.6	55.8	38.4	61.4	22.8	23.8	23.1	23.3
	360	48.6	33.2	59.8	41.7	61.5	25.0	23.9	25.2	23.4
	543	49.6	33.5	61.1	42.8	61.5	25.9	24.0	26.2	23.4
	819	50.1	33.7	61.8	43.4	61.5	26.4	24.0	26.8	23.4
0.8	169	41.5	32.4	52.2	34.8	61.4	21.3	23.8	21.9	23.3
	360	45.3	32.6	55.6	38.2	61.4	22.7	23.8	23.0	23.3
	543	46.8	32.8	57.4	39.8	61.4	23.6	23.9	23.8	23.4
	819	47.7	32.9	58.6	40.7	61.5	24.2	23.9	24.5	23.4
1.485	169	37.8	32.4	50.7	31.9	61.4	20.7	23.8	21.6	23.3
	360	42.0	32.4	52.4	35.1	61.4	21.4	23.8	22.0	23.3
	543	43.9	32.5	54.0	36.8	61.4	22.0	23.8	22.5	23.3
	819	45.2	32.6	55.5	38.2	61.4	22.7	23.8	23.0	23.3

Universal Adhesion Model

The analysis of deviance for the logistic regression model predicting percentage droplet adhesion is shown in Table 5. This tested the significance of velocity, droplet size, velocity x droplet size, leaf angle, leaf surface, and formulation. The deviance for leaf angle has been subdivided into that proportion which can be explained using the sine of the leaf angle, and the remaining deviation from this. In a similar fashion, the deviance for the leaf surface has been subdivided into the portion explained by contact angle and the remaining effect. The deviance for the formulation has been subdivided into the portion explained by (γ_w - γ_t) and the remaining effect. The data show that 91% of the deviance for leaf angle can be explained by the sine of the angle. In the case of leaf surface, 95% of the deviance is explained by the 20% acetone contact angle. Similarly, the dynamic surface tension term (γ_w - γ_t) explains 78% of the differences between solutions, leaving 22% unexplained.

The best model of adhesion is still a logistic regression

$$Adhesion\ (\%\ of\ droplets\ impacting) = 100e^F / (1 + e^F) \quad (8)$$

where

$$F = 16.0 - 6.91(V) + 0.976\ (S) + 2.82\ (V.S) + 4.38(1 - \sin(\theta)) - 0.111\ \varphi_a + 0.0711\ (\gamma_w - \gamma_t) \quad (9)$$

and

V = velocity (m s^{-1}); S = droplet diameter (mm); θ = angle of impact of droplet with leaf surface (degrees); φ_a = acetone contact angle (degrees); γ_w = surface tension of water (72 mN m^{-1}); γ_t = surface tension of the solution at the time of impact.

Table 5 — *Analysis of deviance for the logistic regression model predicting percentage droplet adhesion.*

Source	d.f.	Deviance	% Deviance	Mean Deviance	Deviance Ratio
Velocity	1	7405.2		7405.2	5305.7
Droplet size	1	148.1		148.1	106.1
Velocity x Droplet size	1	398.4		398.4	285.5
Leaf angle	3	3344.2		1114.7	798.7
1-sin(θ)	1	3055.5	91.4	3055.5	2189.2
remainder	2	288.7	8.6	144.4	103.4
Leaf Surface	6	14060.6		2343.4	1679
acetone contact angle, φ_a	1	13304.7	94.6	13304.7	5784.6
remainder	5	755.9	5.4	151.2	65.7
Formulation	10	5677.9		567.8	406.8
(γ_w - γ_t)	1	4431.6	78	4431.6	3175.1
remainder	9	1246.3	22	138.5	99.2
Residual	4327	6039.2		2.3	
Total	4367	37073.6			

P-value < 0.01 for all

Seven coefficients were used in this universal model and the deviance explained was 74.2%, whereas the previous universal model had four coefficients and the deviance explained was 65%.

A specific model for which fall distance, droplet size, angle of impact, leaf surface, and formulation were fitted using individual terms (30 coefficients) provided a percentage deviance explained of 81.4%. Results of this are not shown because a specific model is only useful in showing how well the terms can be modeled. It could not be used for predicting, for example, adhesion using a new formulation. A universal model can be used for predicting the adhesion of, for example, a new formulation, or adhesion to a plant species not already studied.

A test of the universal model showed that the correspondence between calculated and observed mean adhesion values for droplet diameter, fall distance, leaf angle, leaf surface, and formulation was excellent (Table 6).

Table 6 — *Comparison of actual droplet adhesion, and adhesion as predicted by model.*

Factor	Variable	Droplet adhesion (%)	
		Actual	Predicted
Droplet diameter (μm)	169	98.5	98.0
	360	82.5	84.6
	543	71.5	69.1
	819	57.3	58.1
Fall distance (mm)	200	87	85.5
	400	80.8	78.4
	800	75.0	72.8
	1485	64.9	71.0
Leaf angle	0°	87.8	90.1
	22.5°	84.4	81.7
	45°	75.0	71.9
	67.5°	60.4	64.0
Leaf surface[1]	Red Maple, U (70.3°)	94.5	90.7
	Red Maple, L (107.7°)	51.5	57.0
	Sweetgum, U (61.5°)	91.0	94.6
	Sweetgum, L (77°)	84.9	86.6
	Red Oak, U (71°)	93.7	90.3
	Red Oak, L (85.5°)	80.4	80.0
	Lambsquarters, U (123.8°)	42.5	39.1
Formulation and Concentration of triclopyr[2]	1 (3.2 g ae/l)	71.4	75.0
	2 (16 g ae/l)	84.7	81.7
	3 (3.2 g ae/l)	68.8	68.9
	4 (16 g ae/l)	78.7	78.3
	5 (16 g ae/l)	72.6	67.2
	6 (3.2 g ae/l)	83.9	85.1
	7 (16 g ae/l)	86.6	85.4
	8 (3.2 g ae/l)	82.8	85.0
	9 (16 g ae/l)	86.8	85.5
	10	44.9	50.6

[1] U = upper leaf surface, L = lower leaf surface; figures in brackets are contact angle values.
[2] 1and 2 are Garlon® 4; 3 and 4 are Garlon® 3A (commercial formulations of triclopyr); 5 is a base triclopyr formulation; 6 and 7 contain triclopyr + organosilicone surfactant + conventional surfactant; 8 and 9 contain triclopyr + organosilicone surfactant. 10 is water alone.

Despite these improvements, differences between formulations still have not been fully explained. Bergeron *et al.* (1998) studied the adhesion of droplets to a surface in terms of both dynamic surface tension and elongational viscosity. Of the solutions studied, the one whose droplets adhered to the surface had an elongational viscosity two orders of magnitude greater at the higher elongational gradients studied, compared to the solution whose droplets did not adhere. An important point was that for the two solutions whose elongational viscosity's were measured and found to be quite different, they had identical surface tensions (dynamic and equilibrium), thus verifying that their different drop-rebound performance was not due to dynamic surface tension effects. Elongational viscosity needs to be measured for the solutions used in the present study to see if this information can significantly improve the model.

Conclusions

The previous universal adhesion model (Forster et. al. 1998) has been substantially improved. The spray droplet adhesion model using leaf angle, leaf surface and formulation as independent variables has been extended to include droplet size (ranging from 180 μm to 1000 μm) and droplet velocity (0.63 ms^{-1} to 2.82 ms^{-1} at impact). The model has been made more robust by increasing the number of formulations studied (dynamic surface tension of droplets at impact range from 20.7 to 72 $mN\ m^{-1}$), the range of leaf surface types studied (acetone (20%) contact angles on the leaf surfaces ranged from 61.5° to 123.8°) and the range of droplet impact angles studied. In the previous universal adhesion model, water contact angles on the different leaf surfaces accounted for 76% of the differences between leaf surfaces. In this study, contact angles using 20% acetone accounted for 95% of the differences between the leaf surfaces studied. The deviance explained has been increased from 65% to 74%. A test of the universal model showed that the correspondence between calculated and observed mean adhesion values for droplet diameter, fall distance, leaf angle, leaf surface and formulation was excellent.

Acknowledgment

The authors would like to thank Ainsley Robinson for doing the majority of the adhesion measurements. G. Pollicello, OSi Specialties, Crompton Corporation supplied the DST measurements. The New Zealand Public Good Science Fund provided the funding for this work.

References

Bergeron, V., Martin, J-Y., and Vovelle, L., 1998, "Interaction of Droplets with a Surface: Impact and Adhesion," *Proceedings of the Fifth International Symposium on Adjuvants for Agrochemicals*, pp. 377-382.

Brazee, R. D., Reichard, D. L., Bukovac, M. J., and Fox, R. D., 1991, "A Partitioned Energy Transfer Model for Spray Impaction on Plants," *Journal of Agricultural Engineering Research*, Vol. 50, pp. 11-24.

Forster, W. A., and Zabkiewicz, J. A., 1994, "Effect of an Organosilicone Surfactant on Spray Drop Adhesion and Retention by Pea (Pisum Sativum) Leaf Surfaces," *Proceedings of the Forty Seventh New Zealand Plant Protection Conference*, pp. 387-391.

Forster, W. A., Kimberley, M. O., and Zabkiewicz, J. A., 1998, "Spray Droplet Adhesion Models – Evaluation of Critical Solution and Leaf Factors," *Proceedings of the Fifth International Symposium on Adjuvants for Agrochemicals*, pp. 55-60.

Furmidge, C. G. L., 1962, *Journal of Colloid Science,* Vol.17, No 4, pp. 309-324.

Hartley, G. S., and Brunskill, R. T., 1958, "Reflection of water droplets from surfaces," *Surface Phenomena in Chemistry and Biology*, J. F. Danielli; K. G. A. Parkhurst, and A. C. Riddiford, Eds., Pergamon Press, Oxford, pp. 214-226.

Lake, J. R., and Marchant, J. A., 1983, "The Use of Dimensional Analysis in a Study of Drop Retention on Barley," *Pesticide Science*, Vol. 14, pp. 638-644.

McCullagh, P., and Nelder, J.A. 1983, "*Generalised Linear Models,*" Chapman and Hall, New York.

Stevens, P. J. G., Forster, W. A., Murphy, D. S., Policello, G. A., and Murphy, G. J., 1992, Surfactants and Physical Factors Affecting Adhesion of Spray Droplets on Leaf Surfaces," *Proceedings of the Southern Weed Science Society*, Vol. 45, pp. 354-358.

Stevens, P. J. G., Kimberley, M. O., Murphy, D. S., and Policello, G. A., 1993, "Adhesion of Spray Droplets to Foliage: The Role of Dynamic Surface Tension and Advantages of Organosilicone Surfactants," *Pesticide Science*, Vol. 33, pp. 237-245.

Webb, D. A., and Holloway, P. J. 1999, "Low speed impact behaviour of aqueous monosize spray droplets containing agricultural additives on plant surfaces," *Proceedings of the Fifteenth Annual Conference on Liquid Atomisation and Spray Systems*, pp. 1-6.

Webb, D. A., Western, N. M., and Holloway, P. J., 2000, "Modelling the impaction behaviour of agricultural sprays using monosized droplets," *Aspects of Applied Biology* Vol. 57, pp. 147-154.

Zabkiewicz, J. A., 2000, "Adjuvants and herbicidal efficacy – present status and future prospects," *Weed Research,* Vol. 40, No. 1, pp. 139-149.

Paul C. H. Miller,[1] A. J. Hewitt,[2] and W. E. Bagley[3]

Adjuvant Effects on Spray Characteristics and Drift Potential

Reference: Miller, P. C. H., Hewitt, A. J., and Bagley, W. E., "**Adjuvant Effects on Spray Characteristics and Drift Potential**" *Pesticide Formulations and Application Systems: Twenty First Volume, ASTM STP 1414,* American Society for Testing and Materials, West Conshohocken, PA, 2001.

Abstract: This paper describes studies that were conducted to investigate the effects of different surfactant and emulsion adjuvant chemistries and formulations on spray characteristics and drift potential of a commercial herbicide. Previous research has shown that surfactant, emulsion, and polymer adjuvants can affect the formation of sprays and liquid distribution patterns when atomized through flat fan nozzles. The present research involved the application of various different tank mix-adjuvant combinations through flat fan, disc-core, and air induction nozzles. The sprays were characterized using spark photography, droplet size analysis, liquid patternation, and driftability measurements in a wind tunnel. Atomization was related to the physical properties of the tank mixes and sheet breakup mechanisms. The results will help in the development of ASTM test guidelines for spray characterization and drift research, and provide valuable information for applicators to consider when selecting tank mix-nozzle combinations for the application of agricultural chemicals.

Keywords: Spray drift, spray formation, application, adjuvants, spray characterization

Introduction:

An effective agricultural or biological spray application is one where an optimal lethal dose of active ingredient is delivered to the target at minimum rate with minimal environmental loss. Spray drift from the field application of agricultural chemicals is affected by factors related to the equipment (sprayer and nozzles), tank mix physical properties, and meteorological and environmental conditions. The relative importance of these variables may vary from one situation to another. The application parameters of nozzle type, orifice size and spray pressure, and the tank mix physical properties can affect spray performance through determination of the droplet size spectrum that will be formed at atomization. These will also determine the spray pattern (distribution of liquid in the emitted spray plume) and rate of change of droplet size through evaporation. A lot of attention has been given to droplet size effects on spray performance and off-target movements such as drift, with the development of valuable spray classification schemes for spray quality (Anonymous, 1999; Doble et al. 1985). Several researchers (e.g. Miller et al. 1991; Walklate et al. 1994) have shown that droplet characteristics do not affect drift from ground rig applications based on droplet size alone. For some sprays, droplet trajectories, buoyancy, electrostatic charge, and density can also affect movements. Recognizing the importance of drift reduction techniques using low drift nozzles, the International BCPC scheme includes a drift potential assessment (Southcombe et al. 1997). Most assessments using this approach have been limited to equipment assessments such as twin fluid nozzles. The same approach can be used, as shown in the present paper, to assess drift potential for tank mixes containing different adjuvants.

Many previous researchers have observed correlations between the physical properties of tank mixes and the droplet size spectra produced at emission through specific nozzles. The Spray Drift Task Force (SDTF), a consortium of agricultural chemical registrants, conducted approximately 2000 atomization measurements for a near-maximum range of agricultural tank mix surface tension and viscosity physical properties with a wide range of agricultural nozzles and aerial and ground application spray scenarios (Hewitt et al. 1996). The main physical properties that affected droplet size were the dynamic surface tension at a surface lifetime age of 20 millisec., shear viscosity and Effective Trouton Ratio, defined as the maximum inertia-corrected apparent extensional viscosity divided by the shear viscosity. The apparent extensional viscosity was measured over a range of strain rate from 1000-20,000 sec^{-1} (Hewitt 1998, Hewitt and Hermansky 1997). The shear viscosity was measured at a shear rate of 1 sec^{-1}. The SDTF data were used to develop an empirical model, DropKick® that can estimate droplet size spectra and spray quality for aerially applied sprays based on tank mix physical properties and application conditions (nozzle type, orifice size, spray pressure, nozzle angle, and flight speed) for a range of nozzles (Esterly 1998).

With surfactants, dynamic surface tension is particularly important for affecting spray formation and performance (Hewitt and Hermansky 1997). Spanoghe et al. (2000) examined the effects of ethanol and two surfactants ((20POE)-polyoxyethylene sorbitan monolaurate and t-octylphenoxy polyethoxy ethanol) on spray formation and contact angles. For both surfactants at rates up to 1 g/L spray solution, there was no significant effect on the droplet size distribution. On the contrary, when aqueous ethanol solutions were sprayed, a significant decrease in droplet size was found. Furthermore, the droplet size distribution became less uniform when a higher concentration of ethanol in water was used. A decrease in contact angle was found to be more dependent on equilibrium than dynamic surface tension.

Droplet size spectra produced by tank mixes with adjuvants vary with adjuvant and nozzle type (Butler-Ellis and Tuck 1998).

Several researchers have found that the droplet size spectra produced by some nozzle types and tank mix types cannot be related entirely to surface tension and viscosity parameters (Dexter, 1996a, 1996b, 2000, 2001; Butler-Ellis et al. 1999). Dexter (1996a) found that the presence of emulsion particles could affect the sheet breakup mechanism, and Hermansky and Krause (1997) observed that the size of emulsion particles can affect breakup length. Butler-Ellis et al. (1999) studied the breakup of dilute emulsions with and without added surfactants and suggested that the wettability of particles may be correlated with their ability to cause perforations if their interfacial tension is not at equilibrium. It was further suggested that emulsion particles interact with local perturbations within the flow. The addition of surfactants to dilute emulsions caused a reduction in the size and number of emulsion particles, or a change in the number of perturbations, which in turn affected droplet size. The observed trends were for sprays produced by flat fan nozzles. Subsequent work (Butler-Ellis and Tuck 2000) revealed that different nozzle types respond differently to adjuvants. For air induction nozzles, water-soluble surfactants increased and emulsions and dispersions decreased the mean droplet size compared to water alone. Droplet velocity measurements suggested that adding surfactants to water increased the quantity of air included in such droplets; creating an emulsion reduced this quantity.

The purpose of the present study is to extend previous research by investigating the effect of different surfactant and emulsion adjuvants on the important spray efficiency parameters of surface tension, droplet size, liquid distribution patterns, and drift potential. Hewitt and Bagley (2000) reported some preliminary findings from these tests.

Methods

Tank mixes were prepared by adding the following adjuvants to Roundup® Biactive (Monsanto, St. Louis, MO) (glyphosate, 5 % v/v in water) at their normal use rates:

1. silicone surfactant,
2. esterified seed oil and silicone blend,
3. esterified seed oil,
4. paraffinic petroleum oil,
5. nonyl-phenol ethoxylate,
6. surfactant polyblend (aromatic petroleum solvent), and
7. liquid ammonium sulfate/ invert suspension blend.

Each adjuvant was mixed individually with water and glyphosate, N-(phosphonomethyl) glycine, isopropylamine salt at a typical commercial use rate. A water soluble colored dye was added. All components were thoroughly agitated.

The following nozzles (Spraying Systems Co., Wheaton, Illinois) and spray pressures were used for all measurements in this study:

11004 flat fan nozzle at 3 bar pressure,

D5-25 disc-core nozzle at 4 bar pressure, and

11003AI air induction nozzle at 5 bar pressure.

These spray pressures provided matching flow rates of 1.6 L/min for one setting with each nozzle.

Droplet size spectra were characterized using two replicate measurements with a Particle Measuring Systems (PMS, Inc.) 2D optical array probe. To obtain representative cross-section average droplet size spectra measurements, the nozzles were traversed through x and y-axes while sampling at least 10000 droplets per measurement. Droplet size spectra affect spray performance in the field including spray collection on targets, drift minimization and effective dose-transfer to the target end-point (Hewitt 1997).

Given the importance of liquid physical properties on atomization, the dynamic surface tension was measured for each tank mix using a maximum bubble pressure instrument (Fainerman et al. 1994). Measurements were made at surface lifetime ages of 20 millisec. to represent the atomization process of most agricultural sprays.

The sprays were photographed using high-speed photography. Spray angles could be observed from these photographs. Spray angle is an important variable that affects distribution patterns in the field. Breakup length could be assessed to investigate the atomization process for each tank mix/ nozzle combination.

Drift potential was estimated in a low speed wind tunnel using measurement techniques that have been described elsewhere (Walklate et al. 1998). The sprays were emitted in the working section of the wind tunnel with a wind speed of 2.0 m/s. The amount of material airborne as potential airborne drift was measured using sampling heights of 0.1, 0.2, 0.3, 0.4, and 0.5 m heights of strings above the floor of the wind tunnel at a distance of 2.0 m downwind of the nozzle. Sedimentation deposition was estimated by recovery on samplers mounted at a height of 0.1 m at distances of 2, 3, 4, 5, 6 and 7 m downwind of the nozzle. Spray was collected on cylindrical collectors (1.98 mm diameter polythene tubing). These collectors were rinsed into distilled water for colorimetric analysis in the laboratory. Collection rates were expressed as drift in units of percentage of nozzle output of spray. All treatments were replicated to provide two measurements. These measurements provided indications of relative amounts of drift exposure for airborne spray and sedimentation deposition.

Single nozzles were used for patternation of the liquid distribution. A 25 mm fixed trough was used with a technique to monitor flow rates continuously. The instrument waits for a steady flow to be achieved for a given nozzle position, measures the flow by monitoring weight and then moves the nozzle to a new position. A complete liquid distribution pattern is thereby developed on the x and y liquid axes.

Results

The measured downwind horizontal drift profiles was of the same form for all three nozzle conditions examined (Figures 1 to 3).

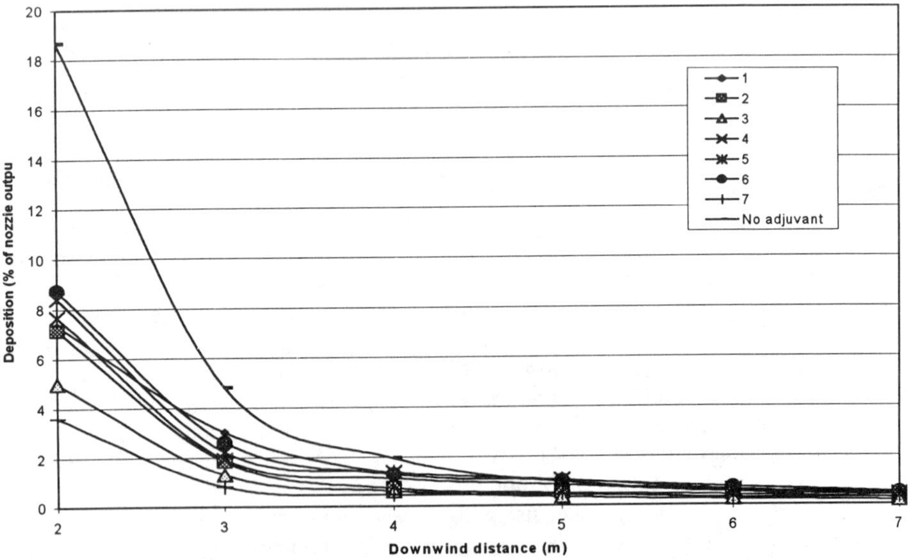

Figure 1. Downwind Deposition for Different Adjuvants With 11004 at 3 bar

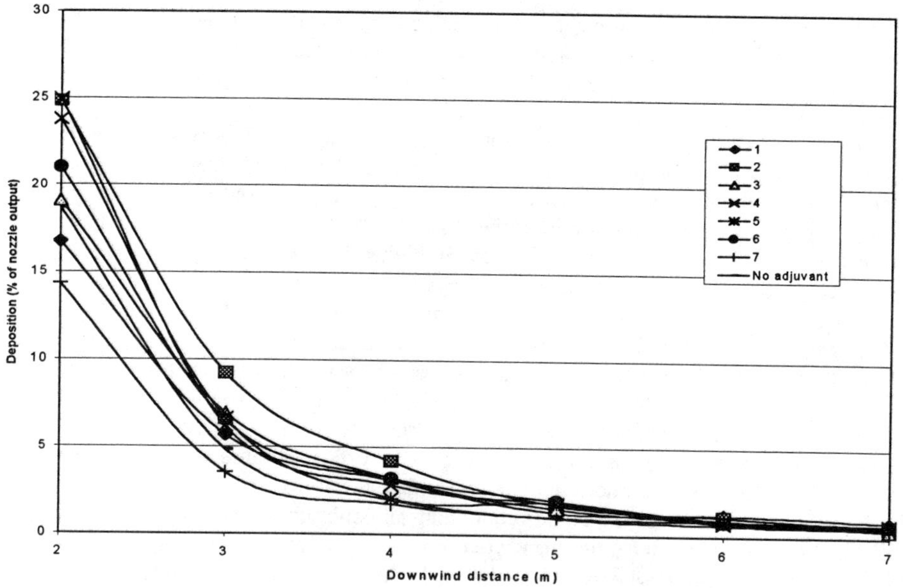

Figure 2. Downwind Deposition for Different Adjuvants With D5-25 at 4 bar

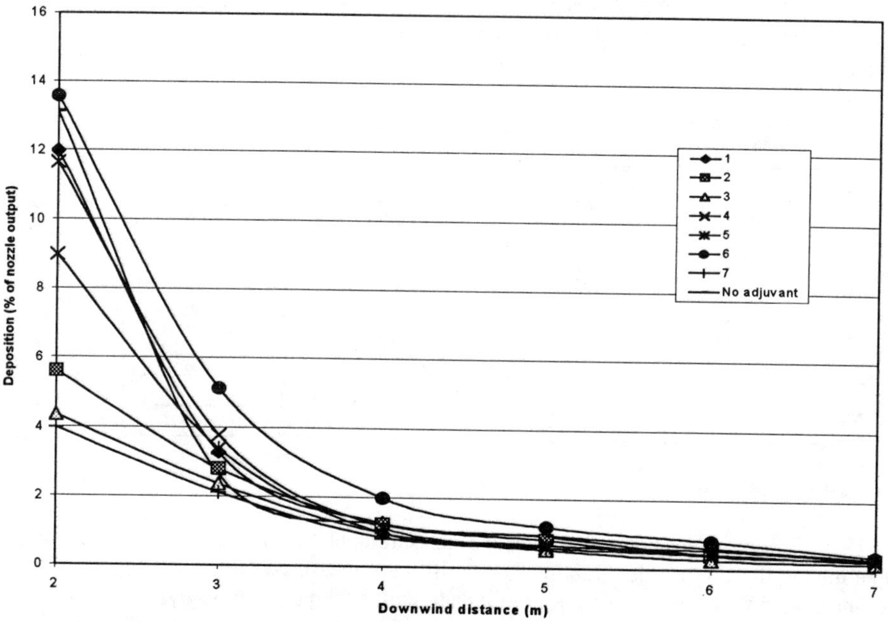

Figure 3. Downwind Deposition for Different Adjuvants With 11003 AI at 5 bar

The results varied with adjuvant and nozzle combinations. In general, liquids 4 and 6 (paraffinic petroleum oil and surfactant polyblend) provided the highest levels of drift with liquid 7 (liquid ammonium sulfate with invert suspension) providing the lowest levels of drift

An example of the vertical airborne spray profile is shown in Figure 4 for the flat fan nozzle. As expected for this nozzle design in which downward droplet velocities are relatively high, the highest levels of airborne spray were measured close to the ground level. Vertical spray profiles for both the disc-core hollow cone and the air induction nozzles gave a higher proportion of spray at higher levels above ground and this is consistent with the reduced velocities from these nozzle designs.

Measurements of the spray volume distribution patterns from the flat fan (Figure 5) and air induction nozzles showed only very small differences when operating with the different spray liquids but larger differences were measured with the disc-core hollow cone nozzle, Figure 6.

Results of measurements of the mean droplet size, expressed as the volume median diameter ($D_{v0.5}$) and the dynamic surface tension at a surface lifetime age of 20 ms are given in Table 1. The differences in dynamic surface tension are relatively small with similar values measured for the liquids containing silicone surfactant and esterified seed oils. The lowest level of dynamic surface tension was measured with the liquid containing the nonyl-phenol ephoxylate and the highest with the liquid containing the surfactant polyblend. Differences in the measured droplet size distributions were also small although it is surprising that the smallest droplet size was measured with the liquid having the highest dynamic surface tension.

The spark photographs provided a visual examination of the spray formation process. Of particular interest was the observation of air bubbles in the liquid sheet from the air induction nozzles.

Table 1. Measured dynamic surface tension and droplet size distributions

Spray liquid	Dynamic surface tension, mN/m	Volume median diameter, µm -flat fan nozzle only
1	53.8	257.2
2	54.2	257.8
3	54.8	259.9
4	55.3	261.6
5	47.8	255.7
6	56.0	248.6
7	50.3	268.6

Discussion and Conclusions

There is some consistency between the measurements of droplet size and spray drift since for the flat fan nozzle, high levels of drift were measured with spray liquids giving the finest sprays. All measurements were made with a formulation of glyphosate and it is likely that the use of a mixture reduces the effects that may have been seen if the adjuvants had been sprayed mixed only with water.

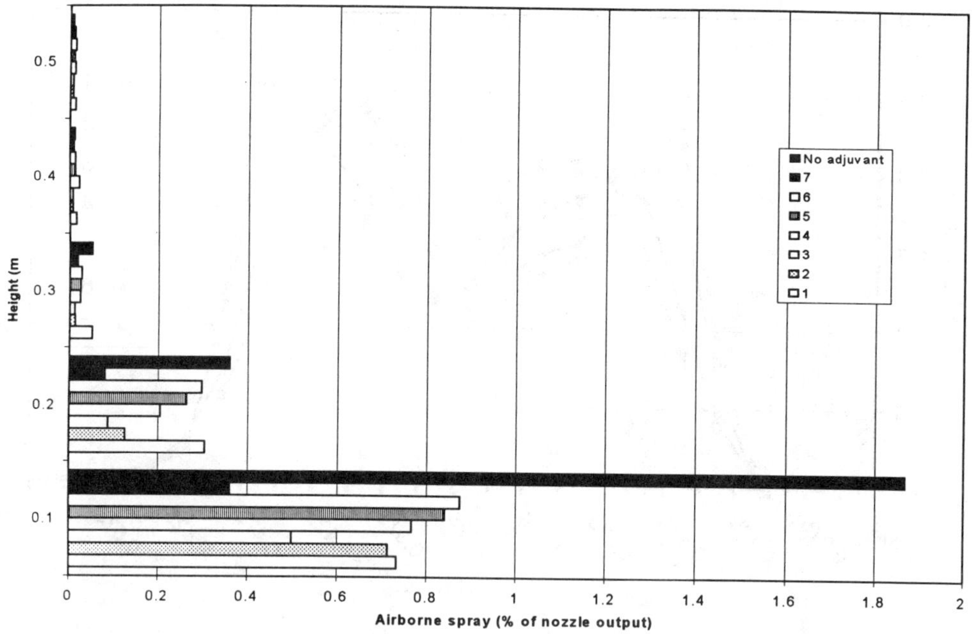

Figure 4. Vertical Airborne Spray Profile for Different Adjuvants With 11004 at 3 bar. Legend Label Order is Same as Data Appear for Each Height

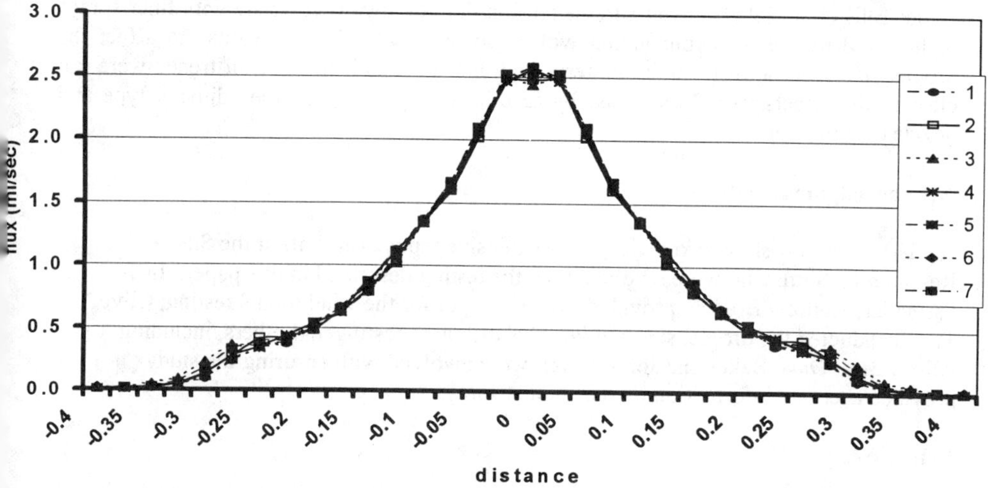

Figure 5. Single flat fan nozzle distribution pattern measured with the different adjuvants

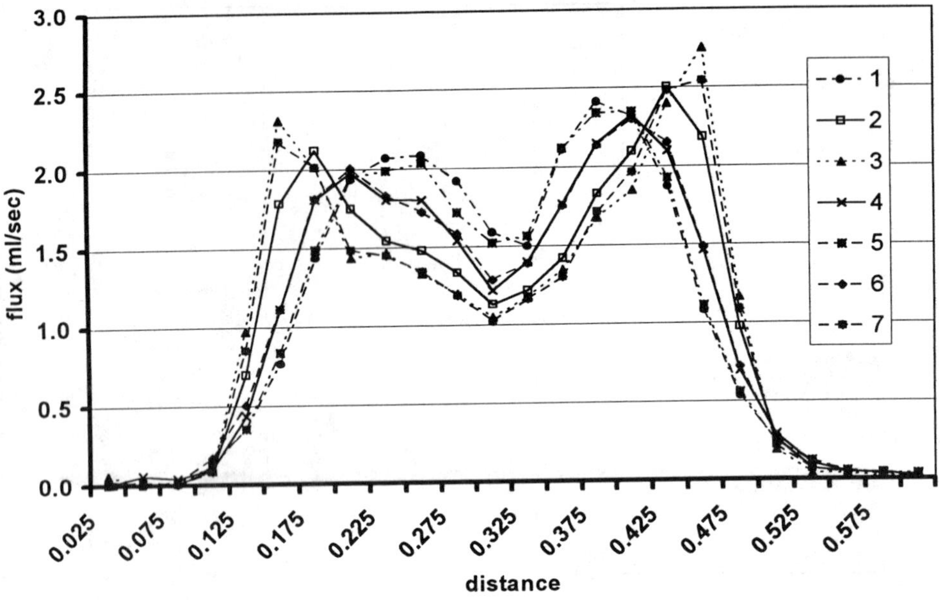

Figure 6. Single disc and core hollow cone nozzle distribution patterns measured with the different adjuvants

In agreement with some previous research (Dexter, 1996a, 1996b, 2000, 2001; Butler-Ellis et al., 1999), lower surface tension did not always correlate with finer sprays or higher drift potential, but varied with adjuvant type. The adjuvants did affect the droplet size spectra, distribution patterns and drift potential, but overall trends were not clear as the effects varied on a case-by-case basis depending on the adjuvant type and nozzle combination.

Acknowledgments

The authors wish to acknowledge the extensive input from staff at the Silsoe Research Institute who were involved with the testing described in this paper. In particular, Andrew Bradley provided essential input for the wind tunnel testing, Clive Tuck conducted the droplet size and liquid distribution testing and others, including Alison Bird, Dave Baker and John Power, were involved with ensuring the study provided accurate information on the applied sprays.

References

Anonymous, 1999, "Spray Nozzle Classification by Droplet Spectra," Standard S-572, American Society for Agricultural Engineers (ASAE), Niles, MI.

Butler-Ellis, M.C. and Tuck, C.R., 1998, "Predicting Spray Droplet Size from Three Nozzle Designs Using Physical Properties of Spray Liquids Containing Adjuvants." *Proceedings 5th International Symposium On Adjuvants and Agrochemicals*, Memphis, TN, pp. 383-400.

Butler-Ellis, M.C., Tuck, C.R. and Miller, P.C.H., 1999, "Dilute Emulsions and their Effect on the Breakup of the Liquid Sheet Produced by Flat-Fan Spray Nozzles," *Atomization and Sprays*, Vol. 9, pp. 385-97.

Butler-Ellis, M.C. and Tuck, C.R., 2000, "The Variation in Characteristics of Air-Included Sprays with Adjuvants," *Aspects of Applied Biology*, Vol. 57, pp. 155-62.

Dexter, R.W., 1996a, "Measurement of Extensional Viscosity for Polymer Solutions and its Effects on Atomization from a Spray Nozzle," *Atomization and Sprays* Vol. 6, pp. 167-91.

Dexter, R.W., 1996b, "Interactions of Anionic Surfactants and Polymers used as Spray Tank Adjuvants," *Pesticide Formulation and Application Systems: 16th Volume ASTM STP 1312*, M. J. Hopkinson, H. M. Collins, and G. R. Goss, Eds, American Society for Testing and Materials, Philadelphia, PA.

Dexter, R.W., 2000, "Emulsion Properties and Agricultural Spray Quality," *Proceedings Eighth International Conference on Liquid Atomization and Spray Systems*, Pasadena, CA, pp. 341-8.

Dexter, R., 2001, "Fluid Properties and Spray Quality," *Pesticide Formulations and Applications Systems: Twentieth Volume, ASTM STP 1400*, A. K. Viets, R. S. Tann and J. C. Mueninghoff, Eds., American Society for Testing and Materials, West Conshohocken, PA.

Doble, S. J., Matthews, G. A., Rutherford, I., and Southcombe, E. S. E., 1985, "A System for Classifying Hydraulic Nozzles and Other Atomizers into Categories of Spray Quality," *Proceedings of 1985 British Crop Protection Conference – Weeds*, Vol. 9A-6, pp. 1125-33.

Esterly, D M., 1998, "Neural Network Analysis of Spray Drift Task Force DROPKICK II™," *ASAE Paper No. 981014*, Orlando, FL, American Society for Agricultural Engineers, Niles, MI.

Fainerman, V.B., Miller, R. and Joos, P., 1994, "The Measurement of Dynamic Surface Tension by the Maximum Bubble Pressure Method," *Colloid Polymer Science*, Vol. 272, pp. 731.

Hermansky, C. G. and Krause, G. F., 1997, "Relevant Physical Property Measurements for Adjuvants," *Proceedings ISAA 4th International Conference*, pp. 20-2.

Hewitt, A. J., 1997, "The Importance of Droplet Size in Agricultural Spraying," *Atomization & Sprays,* Vol. 7, No. 3, pp. 235-244.

Hewitt, A.J., 1998, "The Effects of Tank Mix and Adjuvants on Spray Drift," *Proceedings of the Fifth International Symposium on Adjuvants for Agrochemicals*, Memphis, TN.

Hewitt, A.J. and Bagley, W.E., 2000, "Evaluation of the Effects of Adjuvants on Agricultural Spray Characteristics," *Proceedings of the Eighth International Conference on Liquid Atomization and Spray Systems*, Pasadena, CA, pp. 386-90.

Hewitt, A.J. and Hermansky, C.G., 1997, "Effect of Liquid Properties on Spray Performance," *Proceedings Chem Show*, New York, NY.

Hewitt, A. J., Valcore, D. L. and Bryant, J. E., 1996, "Spray Drift Task Force Atomization Droplet Size Spectra Measurements," *Proceedings ILASS-Americas 96*, San Francisco, CA.

Miller, P.C.H., Tuck, C.R., Gilbert, A.J. and Bell, G.J., 1991, "The Performance Characteristics of a Twin-Fluid Nozzle Sprayer," *BCPC Monograph No. 46, Air-Assisted Spraying in Crop Protection*, British Crop Protection Council, Swansea, UK, pp. 97-106.

Southcombe, E. S. E., Miller, P. C. H., Ganzelmeier, H., Van de Zande, J. C., Miralles, A., and Hewitt, A. J., 1997, "The International (BCPC) Spray Classification System Including a Drift Potential Factor," *Proceedings British Crop Protection Conference,* Vol. 5A-1, pp. 371-380.

Spanoghe, P., Van der Meeren, P. and Steurbaut, W., 2000, "The Influence of Dynamic Surface Tension on Atomisation and Retention of Agrochemical Active Ingredients," *Proceedings 5th World surfactants Congress, Cesio*, Committee European des Agents de Surface et Leurs Intermediaires Organiques, Vol. 2, pp. 899-907.

Walklate, P.J., Miller, P.C.H., Rubbis, M. and Tuck, C.R., 1994, Agricultural Nozzle Design for Spray Drift Reduction, *Proceedings ICLASS-94*, Rouen, France, pp. 851-858.

Walklate, P.J., Miller, P.C.H., Richardson, G.M. and Baker, D.E., 1998, "A Similarity Scaling Principle for Risk Assessment of Spray Drift" *Proceedings of the 14th Conference on Liquid Atomisation and Spray Systems, ILASS Europe 98*, pp. 499 – 504.

Masoud Salyani,[1] Ezzeddine BenSalem,[1] and Jodie D. Whitney[1]

Spray Deposition Effect on Abscission Efficacy of CMN-Pyrazole in Harvesting Oranges

---

Reference: Salyani, M., BenSalem, E., and Whitney, J. D., **"Spray Deposition Effect on Abscission Efficacy of CMN-Pyrazole in Harvesting Oranges,"** *Pesticide Formulations and Application Systems: A New Century for Agricultural Formulations, Twenty First Volume, ASTM STP 1414,* J. C. Mueninghoff, A. K. Viets, R. A. Downer, Eds., American Society for Testing and Materials, West Conshohocken, PA, 2001.

Abstract: The objective of this research was to determine the relationship between spray deposition and abscission efficacy of (CMN-Pyrazole), in mechanical harvesting of the 'Hamlin' orange. Two tests were conducted on 1/6/00 and 2/11/00 using an air-blast tower sprayer. The sprayer was equipped with an axial-flow fan and hydraulic nozzles. The trees were sprayed at three nozzle arrangements so that 3/4, 1/2, or 1/4 of the total discharge was directed to the upper half of the tree canopy and the rest to the lower half. Spray mixtures, containing CMN-Pyrazole (Release®), adjuvant Kinetic, and a fluorescent tracer, were applied at 2.6 km/h, 3,750 L/ha and 4.0 km/h, 2,450 L/ha, in the first and second tests, respectively. Leaf samples were collected from different canopy locations and spray deposition was quantified by fluorometry. Four to five days after the applications, fruit detachment force (FDF) was measured, dropped fruit was counted, the trees were shaken with a trunk shaker, and percentage of fruit removal was determined.

Mean canopy deposition was similar for all treatments. Deposition at different canopy heights was not proportional to the amount of spray directed to that height. The situation was improved in the second experiment, probably due to the lower application rate and reduced runoff. There was no difference in deposition between two sides of the canopy. Outside canopy deposition was higher than inner canopy deposition in most locations. In general, spray applications were effective in reducing FDF. There was stronger correlation between spray deposition and FDF in the second test. FDF was higher in the upper canopy compared to lower locations in the unsprayed control trees. Fruit removal efficiencies of sprayed trees were significantly higher than those of the unsprayed control in the second test.

Keywords: citrus, growth regulators, spray droplet, fruit detachment force, shaker

---

[1]Associate Professor, Research Associate, and Professor of Agricultural Engineering, respectively, University of Florida, IFAS, Citrus Research and Education Center, 700 Experiment Station Road, Lake Alfred, FL 33850, USA.

A key requirement for feasibility of mechanical harvesting of Florida citrus is high fruit removal efficiency. Without abscission chemicals, fruit removal efficiencies of the existing mechanical harvesters range between 70 and 95% (Whitney 1999). The percentage of fruit remaining on the tree is too large to be ignored and gleaning of the tree by manual labor is usually not an economical option either. An effective abscission agent to reduce the fruit detachment force could raise the fruit removal efficiency of the harvesters to an acceptable level. It could also reduce the need for aggressive shaking of the tree, thus decrease potential wear of the equipment and minimize injuries to the trunk and root system.

Wilson et al. (1981) investigated the abscission potential of several growth regulators. Some proved effective but each had certain limitations and none was found suitable for further development and registration. Hedden et al. (1988) found that abscission chemicals could increase fruit removal efficiencies of the trunk shakers by 17 to 26 percentage points. Whitney et al. (1999) observed 50 to 80% decrease in fruit detachment force (FDF), using prosulfuron, metsulfuron-methyl, and CMN-pyrazole as abscission agents. Nonetheless, the efficiencies of the chemicals were not consistent and depended on the tree cultivar, applied dosage, and timing of the applications.

Prosulfuron, applied at different times, was found effective in loosening 'Hamlin' (early season cultivar) but the response was erratic for the late season 'Valencia' oranges (Kender et al. 1998). Metsulfuron-methyl was also effective in reducing FDF and shaker efficiency in 'Hamlin,' but heavy defoliation and twig dieback were observed at high doses and volume rates (Kender et al. 1999b). Twig dieback was also observed in 'Marsh' grapefruit (Kender et al. 1999a). These phytotoxic responses were not reported earlier by Wilcox and Taylor (1997). Hartmond et al. (2000), applying metsulfuron-methyl and CMN-pyrazole to 'Valencia' orange at 10-day intervals, observed a "less-responsive period" following bloom. They also reported undesirable side effects for metsulfuron-methyl. CMN-pyrazole was not phytotoxic and showed much faster response. Considering its efficacy, phytotoxicity, and safety, CMN-pyrazole has been identified as the most desirable experimental abscission compound for loosening citrus fruit under Florida conditions. Therefore, several field experiments have been conducted to define the desirable application parameters for different conditions.

Koo et al. (1999, 2000) investigated the relationship between spray variables and fruit removal efficiency of a trunk shaker in 'Hamlin' and 'Valencia' oranges. Deposition decreased with spray volume but higher volume and CMN-pyrazole rates resulted in lower fruit detachment forces and higher fruit removals than lower rates. Increasing the sprayer ground speed at constant sprayer output improved the deposition efficiency. However, more uniform spray distribution, better canopy penetration, and higher fruit removal efficiencies were obtained at lower ground speeds. BenSalem et al. (2000) found higher deposition and fruit removal efficiencies from night-time compared to day-time applications.

The objective of this study was to determine the relationship between spray deposition and abscission efficacy of CMN-pyrazole (CMN-P), in mechanical harvesting of 'Hamlin' orange.

Materials and Methods

Spray Applications

An experimental fruit abscission agent, CMN-pyrazole (5-chloro-3-methyl-4-nitro-1-H-pyrazole), was applied to 'Hamlin' orange trees (Citrus sinensis (L.) Osbeck) on 1/6/00 (Test 1) and 2/11/00 (Test 2). The trees were 5.5 to 6.5 m high and spaced at 3.7 by 7.6 m, within and between the rows, respectively. The formulated product, Release® (17.18% a.i. w/w), was developed by Abbott Laboratories (Chicago, IL), in the early 1970s. Spray mixtures contained Release® at 100 mg a.i./L, adjuvant 'Kinetic' at 0.1% v/v, and a deposition tracer. The tracers were Yellow Fluorescent Pigment EC (SARDI, Loxton, Australia) at 1.5 mL/L (Test 1) and Pyranine 10G (Keystone Aniline Corp., Chicago, IL) at 350 mg/L (Test 2). The mixtures were applied at 2.6 km/h, 3750 L/ha and 4.0 km/h, 2450 L/ha, in the first and second tests, respectively.

The sprayer was an engine-driven air-blast (Titan1093, John Bean Sprayers, Hogansville, GA) equipped with an axial-flow fan, a tower attachment, and 34 disc-core nozzles (at about 15 cm spacings) per side. It was operated at three nozzle arrangements: T3B1 (¾-¼), T2B2 (½-½), and T1B3 (¼-¾) so that ¾, ½, or ¼ of the total spray flow was discharged from the upper (top) 17 nozzles and directed to the upper half of the tree canopy. The rest of the flow was discharged from the lower (bottom) 17 nozzles and directed to the lower part of the canopy (Table 1). The three spray treatments and the unsprayed control (check) were replicated four times and assigned to 16 three-tree plots, in randomized complete block design. After spraying, leaf samples were collected from two sides of the canopy at three heights (1.5, 3.0, and 4.5 m) and two canopy depths (on the outer periphery (outside) and 0.6-0.8 m inside). The samples (four to five per location) were placed in sealable plastic bags and stored in an ice chest/refrigerator until they were analyzed, three to seven days later. The samples were washed with de-ionized water and the tracer deposits were quantified by fluorometry (Salyani 2000a). The surface areas of the sample leaves were then measured by an area meter (Delta-T Devices, Burwell, England). The tracer deposition data were used to calculate CMN-P deposition.

Table 1- *Nozzle arrangement of the spray treatments.*

Treatment Code[1]	Nozzles Upper 17	Nozzles Lower 17	Nozzle Pressure kPa	Total Flow Rate[3], L/min Upper 17	Total Flow Rate[3], L/min Lower 17
T3B1 (¾-¼)	D6-25[2]	D4-23	807	47.1	15.7
T2B2 (½-½)	D4-25	D4-25	1076	31.4	31.4
T1B3 (¼-¾)	D4-23	D6-25	807	15.7	47.1
Check	-	-	-	0	0

[1]All treatments were applied with John Bean Titan 1093 sprayer.

[2] Ceramic disc-core nozzles (Spraying Systems, Co., Wheaton, IL).

[3] The flow rates were measured by the tank refill method.

Harvesting Measurements

Four days after spray applications, fruit detachment force (FDF) was measured with a pull force gauge (FDV-50, Wagner Instrument, Greenwich, CT). Using the middle tree in each plot, measurements were made on each of five fruit located at about 1.5 m and 4.5 m high. The CMN-pyrazole induced fruit drop was counted and converted to fruit weight/tree (based on average fruit weight of detached samples).

Five days post spray, the center trees of each plot were shaken with an Orchard Rite (Yakima, WA) trunk shaker for about 10 s, using a multidirectional pattern (Whitney et al. 2000). In Tests 1 and 2, the shaker unbalanced masses were 250 and 205 kg at 14.7 cm eccentricity, respectively. Shaker clamp height on the trunk and the trunk circumference were measured on each tree. Shaker-tree motion and frequency were determined from accelerometer (PCB Piezotronics, Inc., Depew, NY) readings on representative trees. The removed fruit (including fruit drop) and gleaned fruit (fruit left on tree after shaking) were weighed and fruit removal efficiency of the treatments calculated.

Average minimum and maximum temperatures between spray and harvest were 15.6-26.7°C (Test 1) and 12.2-25.6°C (Test 2) at a weather station within 5 km of the test site.

Statistical Analyses

Spray deposition data, expressed in ηg a.i./cm^2, were analyzed by the mixed model procedure (PROC MIXED, with LSMEANS and PDIFF options) (Littell et al. 1996). The means were separated by LSD at the 5% level. Mechanical harvesting data (fruit detachment force, percent removal, yield, etc., were analyzed by the general linear model (GLM) procedure and the means were separated by Duncan's multiple range test at the 5% level (SAS 1990). Variability of data was expressed in terms of the coefficient of variation (CV). The relationships between spray deposition and harvesting parameters were explained by the Pearson correlation coefficient (SAS 1990).

Results and Discussion

Spray Deposition

Test 1 - Mean canopy deposition was similar for all treatments (Table 2); however, T3B1 nozzle arrangement appeared to provide more uniform deposition (CV=22%) compared to other treatments (27 and 38%). Mean deposition at different canopy heights were not proportional to the sprayer discharge rates at different levels (¾-¼, ½-½, or ¼-¾) but followed the expected trend to some extent. With T1B3, the lowest and highest depositions were achieved at the 4.5 and 1.5 m heights, respectively. The trend was reversed for T3B1 and with T2B2; deposition was about the same along the canopy height. Comparatively lower depositions at the upper canopy locations may be attributed to the 'shingling' of the leaves, spray runoff from the upper to the lower levels, and more droplet drift from the upper nozzles. Overall, the outside deposition was higher than the inside deposition for all treatments. However, at the 1.5 m height, where the tree canopy was thinner and nearer to the sprayer, outside deposition was less than inside deposition (in T1B3 and T3B1). This result could in part be attributed to higher sprayer air velocity and excessive runoff at that level (Salyani et al. 2000). The trend was reversed at the 4.5 m level. There was no difference in deposition between two sides of the canopy (data not shown).

Table 2 - *Mean CMN-pyrazole deposition in Test 1.*

Treat. Code	Sample Height m	Deposition Mean (CV) Outside ηg /cm^2	Inside ηg /cm^2	Mean ηg /cm^2	Deposition Mean (CV) Outside ηg /cm^2	Inside ηg /cm^2	Mean ηg /cm^2
T1B3	4.5	177.1 a[1]	127.9 b	152.5 b			
		(36)	(40)	(40)			
	3.0	166.3 a	168.4 ab	167.3 ab	175.8 a A	160.9 a A	168.3 a
		(36)	(36)	(35)	(33)	(45)	(38)
	1.5	183.9 a	186.5 a	185.2 a			
		(30)	(50)	(40)			
T2B2	4.5	215.3 a	120.1 b	167.7 a			
		(16)	(36)	(37)			
	3.0	178.0 ab	147.3 ab	162.6 a	188.1 a A	144.6 a B	166.3 a
		(22)	(17)	(22)	(20)	(28)	(27)
	1.5	171.0 b	166.4 a	168.7 a			
		(15)	(24)	(19)			
T3B1	4.5	202.1 a	151.4 a	176.8 a			
		(16)	(16)	(22)			
	3.0	164.5 ab	159.8 a	162.1 a	172.0 a A	159.1 a A	165.6 a
		(21)	(28)	(24)	(23)	(21)	(22)
	1.5	149.5 b	166.3 a	157.9 a			
		(24)	(18)	(21)			
					Outside	Inside	
			Mean Deposition		178.6 A	154.9 B	
			(CV)		(26)	(33)	

[1]Means followed by the same lower case letter(s) in each column are not significantly different at the 5 % level. Upper case letters compare the outside and inside depositions. CV is the coefficient of variation of the means in %.

Test 2 - There was no significant difference in mean canopy deposition of different treatments (Table 3). Similar to Test 1, T3B1 provided more uniform deposition (CV=26%) compared to other treatments (39 and 51%). Deposition at different canopy heights were significantly different. At 1.5 m, the highest and lowest depositions were observed for T1B3 and T3B1, respectively. At the 4.5 m height, there was an opposite trend. T2B2 showed intermediate values. Although there was clearer separation of depositions at different heights for various treatments, the results did not follow the expected proportions (¾-¼, ½-½, or ¼-¾) in nozzle discharge rates. These results are in agreement with those of Koo et al. (2000) and Salyani et al. (2000) who observed lower depositions in the upper parts of the canopy, using the same sprayer in uniform nozzle arrangement (½-½). In most outside locations depositions were higher than inside; however, the means were not significantly different for any treatment. Again, T1B3 nozzle arrangement gave higher inside deposition than outside at the lowest canopy location. Inner canopy depositions were more variable (higher CVs) than outside. There was no difference in deposition between two sides of the canopy (data not shown).

Table 3 - *Mean CMN-pyrazole deposition in Test 2.*

Treat. Code	Sample Height (m)	Deposition Mean (CV) Outside (ηg /cm^2)	Deposition Mean (CV) Inside (ηg /cm^2)	Deposition Mean (CV) Mean (ηg /cm^2)	Deposition Mean (CV) Outside (ηg /cm^2)	Deposition Mean (CV) Inside (ηg /cm^2)	Deposition Mean (CV) Mean (ηg /cm^2)
T1B3	4.5	82.1 c[1] (22)	57.5 b (21)	69.8 c (28)	123.5 ab A (33)	109.1 a A (68)	116.3 a (51)
	3.0	122.8 b (23)	72.0 b (46)	97.4 b (41)			
	1.5	165.5 a (13)	197.6 a (28)	181.6 a (24)			
T2B2	4.5	77.8 b (21)	80.4 b (51)	79.1 b (38)	111.8 b A (38)	108.9 a A (41)	110.4 a (39)
	3.0	99.0 b (13)	92.4 b (24)	95.7 b (19)			
	1.5	158.8 a (23)	154.0 a (17)	156.4 a (21)			
T3B1	4.5	116.5 b (21)	94.9 b (31)	105.7 b (27)	136.7 a A (21)	117.4 a A (31)	127.1 a (26)
	3.0	128.9 b (13)	122.6 ab (18)	125.8 b (15)			
	1.5	164.6 a (10)	134.8 a (33)	149.7 a (24)			
			Mean Deposition (CV)		Outside 124.0 A (31)	Inside 111.8 A (48)	

[1]Means followed by the same lower case letter(s) in each column are not significantly different at the 5 % level. Upper case letters compare the outside and inside depositions. CV is the coefficient of variation of the means in %.

Comparison of mean depositions in Test 1 and Test 2 shows lower depositions for the latter. This result is due to the increase of sprayer ground speed from 2.6 to 4.0 km/h and consequent reduction of application rate from 3750 to 2450 L/ha, in the second test. However, for the same CMN-P output rate (sprayer discharge per unit travel distance), higher deposition efficiencies (spray deposit per unit travel distance) were achieved in the second test. This result agrees with those of Salyani and Whitney (1990), Salyani (2000b), and Koo et al. (2000) and may partially be attributed to the reduction spray runoff from leaf surface at lower application rate. Overall, the results showed good spray penetration for all treatments.

Mechanical Harvesting

Test 1 - The results are summarized in Table 4. At harvest, the Brix/acid ratio (Ratio) averaged 17.0. The degree Brix is percent soluble solids (sugar) in orange juice and the 'Ratio' is a measure of fruit maturity and juice quality. The fruit detachment force (FDF) in the lower canopy was significantly reduced only by T1B3, whereas in the upper canopy, all CMN-P treatments reduced FDF significantly. The FDF of unsprayed (check) trees was significantly higher in the upper canopy than in the lower part of the tree. These results are in agreement with those of Kender and Hartmond (1999). Fruit drop was minimal and increased by abscission sprays. The trees were uniform in that there were no significant differences among trunk circumferences and fruit yields. Shaker clamp heights were not significantly different among treatments. Fruit removal (drop + shaker removal) was not significantly affected by spray treatments. Shaker displacement and frequency were 7 cm and 7 Hz.

Test 2 - The results are summarized in Table 5. The FDF of check trees dropped to 61.1 N compared to 79.8 N in Test 1, probably due to the increase in fruit maturity (the Brix/acid ratio averaged 18.3). Only the T3B1 treatment significantly reduced the FDF in upper levels. Fruit drop was significantly increased by all spray treatments. Again, the trees were similar in trunk circumference and yield. Shaker clamp heights on the trunk were significantly higher for T2B2 and T3B1. All spray treatments significantly increased fruit removal, with an overall average increase of about 10 percentage points. Shaker displacement and frequency were 6 cm and 7 Hz.

Correlations among Variables

Test 1 - The Pearson correlation coefficients of the test variables are shown in Table 6. CMN-P spray deposition was positively correlated with fruit drop. There was a negative correlation between deposition and fruit detachment force (FDF). The correlation was significant for the upper canopy FDF. Fruit removal also did not appear to have meaningful correlations with spray deposition, clamp height, trunk circumference, or fruit yield.

Test 2 - The Pearson correlation coefficients of the test variables are shown in Table 7. The upper and overall FDFs showed significant negative correlations with spray deposition. The latter was also significantly correlated with fruit drop and removal at all levels. The upper and overall FDFs were also negatively correlated with fruit drop. Fruit removal showed significant correlations with the shaker clamp height (+) and trunk circumference (-).

Table 4 - *Fruit harvesting parameters in Test 1.*

Treat. code	Fruit Detachment Force Upper (N)	Lower (N)	Mean (N)	Clamp Height (cm)	Trunk Circumf. (cm)	Fruit Drop (%)	Fruit Removal (%)	Fruit Yield/tree (kg)
T1B3	66.2 b A[1]	53.6 b A	59.9 b	20.6 a	78.1 a	1.2 ab	92.1 a	228.0 a
	(35)	(29)	(30)	(15)	(2)	(35)	(4)	(7)
T2B2	60.7 b A	64.1 ab A	62.4 b	23.5 a	73.0 a	1.4 ab	91.3 a	231.8 a
	(25)	(17)	(21)	(24)	(6)	(32)	(4)	(9)
T3B1	61.7 b A	66.3 ab A	64.0 b	22.9 a	74.0 a	1.5 a	94.8 a	242.2 a
	(18)	(9)	(12)	(30)	(10)	(36)	(4)	(7)
Check	87.3 a A	72.3 a B	79.8 a	22.2 a	71.8 a	0.6 b	93.0 a	230.2 a
	(13)	(5)	(7)	(7)	(7)	(82)	(4)	(24)

[1] Means followed by the same lower case letter(s) in each column are not significantly different at the 5 % level. Upper case letters compare the upper and lower FDF for each treatment. CV is the coefficient of variation of the means in %.

Table 5 - *Fruit harvesting parameters in Test 2.*

Treat. code	Fruit Detachment Force Upper (N)	Lower (N)	Mean (N)	Clamp Height (cm)	Trunk Circumf. (cm)	Fruit Drop (%)	Fruit Removal (%)	Fruit Yield/tree (kg)
T1B3	55.5 a A[1]	46.4 a A	51.0 ab	23.5 ab	77.2 a	1.7 bc	96.7 a	223.2 a
	(16)	(14)	(11)	(22)	(1)	(38)	(1)	(3)
T2B2	48.3 ab A	42.0 a A	45.2 b	26.1 a	69.6 a	1.9 ab	95.8 a	203.3 a
	(39)	(31)	(31)	(12)	(9)	(30)	(2)	(14)
T3B1	33.6 b B	51.6 a A	42.6 b	18.1 b	74.3 a	2.9 a	93.4 a	234.0 a
	(22)	(38)	(27)	(9)	(8)	(29)	(1)	(14)
Check	64.5 a A	57.7 a A	61.1 a	19.7 b	75.6 a	0.7 c	84.9 b	228.2 a
	(5)	(14)	(8)	(19)	(21)	(51)	(5)	(7)

[1]Means followed by the same lower case letter(s) in each column are not significantly different at the 5 % level. Upper case letters compare the upper and lower FDF for each treatment. CV is the coefficient of variation of the means in %.

Table 6 - *Pearson Correlation Coefficients (r (P>r)) in Test 1.*

Variable	Canopy Location	Fruit Detachment Force Upper	Fruit Detachment Force Lower	Fruit Detachment Force Mean	Fruit Drop	Fruit Removal
CMN-P Deposition	Upper	-0.54 (0.03)			0.65 (0.01)	-0.15 (0.57)
	Lower		-0.40 (0.12)		0.55 (0.03)	-0.20 (0.46)
	Overall			-0.48 (0.06)	0.62 (0.01)	-0.18 (0.50)
Fruit Detachment Force	Upper				-0.34 (0.20)	-0.22 (0.41)
	Lower				-0.09 (0.73)	-0.18 (0.51)
	Overall				-0.27 (0.31)	-0.23 (0.39)
Clamp Height						0.12 (0.65)
Trunk Circumference						-0.10 (0.72)
Fruit Yield (per tree)						-0.23 (0.38)

Conclusions

1. Nozzle arrangement had a significant effect on spray deposition; however, the magnitudes of deposition at different heights were not proportional to the sprayer discharge rates.

2. Spray applications were effective in reducing the fruit detachment force, particularly at the upper part of the canopy.

3. There were good correlations between spray deposition at different levels and fruit drop and removal, in the second test (probably aided by the increased fruit maturity).

4. Fruit removal was positively correlated with the shaker clamp height and negatively with the trunk circumference.

Table 7 - *Pearson Correlation Coefficient (r (P>r)) in Test 2.*

Variable	Canopy Location	Fruit Detachment Force Upper	Fruit Detachment Force Lower	Fruit Detachment Force Mean	Fruit Drop	Fruit Removal
CMN-P Deposition	Upper	-0.71 (0.002)			0.78 (0.0004)	0.68 (0.004)
	Lower		-0.30 (0.25)		0.55 (0.03)	0.89 (0.0001)
	Overall			-0.58 (0.02)	0.68 (0.004)	0.84 (0.0001)
Fruit Detachment Force	Upper				-0.55 (0.03)	-0.27 (0.31)
	Lower				-0.37 (0.16)	-0.37 (0.15)
	Overall				-0.58 (0.02)	-0.39 (0.13)
Clamp Height						0.43 (0.09)
Trunk Circumference						-0.43 (0.09)
Fruit Yield (per tree)						-0.34 (0.20)

Acknowledgments

Florida Agricultural Experiment Station Journal Series No. R-07868.

The research was in part supported by a grant from the Florida Department of Citrus, Citrus Harvesting Council.

The mention of product and company names is for information only and does not imply an endorsement or recommendation by the University of Florida.

The authors wish to thank Roy D. Sweeb, David Noxel, and Gerald Perkins for their technical assistance in conducting the experiments.

References

BenSalem, E., Salyani, M., and Whitney, J. D., 2000, "Spray Deposition and Abscission Efficiency of CMN-pyrazole Used to Enhance Mechanical Harvesting of 'Valencia' Orange," submitted for publication in *Applied Engineering in Agriculture*.

Hartmond, U., Whitney, J. D., Burns, J. K., and Kender, W. J., 2000, "Seasonal Variation in the Response of 'Valencia' Oranges to Two Abscission Compounds," *HortScience,* Vol.35, No. 2, pp. 226-229.

Hedden, S. L., Churchill, D. B., and Whitney, J. D., 1988, "Trunk Shakers for Citrus Harvesting - Part II: Tree Growth, Fruit Yield, and Removal," *Applied Engineering in Agriculture*, Vol. 4, No. 2, pp.102-106.

Kender, W. J. and Hartmond, U., 1999, "Variability in Detachment Force and Other Properties of Fruit Within Orange Tree Canopies," *Fruit Varieties Journal,* Vol. 53, No. 2, pp.105-109.

Kender, W. J., Hartmond, U., Salyani, M., Burns J. K., and Whitney, J. D., 1998, "Effects of Concentration and Application time of 'Prosulfuron' on the Abscission of 'Hamlin' and 'Valencia' Oranges," *Proceedings of the Florida State Horticultural Society*, Vol. 111, pp.108-112.

Kender, W. J., Hartmond, U.,and Burns J. K., 1999a, "Fruit Abscission and Leaf Drop in Citrus Cultivars Treated with Metsulfuron-methyl," *HortTechnology,* Vol. 9, No. 3, pp. 412-416.

Kender, W. J., Hartmond, U., Salyani, M., Burns J. K., and Whitney, J. D., 1999b, "Injurious effects of metsulfuron-methyl when used to stimulate abscission of 'Hamlin' oranges," *HortScience*, Vol. 34, No. 5, pp. 907-910.

Koo, Y. M., Salyani, M., and Whitney, J. D., 1999, "Effects of Abscission Chemical Spray Deposition on Mechanical Harvest Efficacy of 'Hamlin' Orange," *Proceedings of the Florida State Horticultural Society*, Vol. 112, pp. 28-33.

Koo, Y. M., Salyani, M., and Whitney, J. D., 2000, "Spray Variable Effects on Abscission of Orange Fruit for Mechanical Harvesting," *Transactions of the ASAE*, Vol. 43, No. 5, pp.1067-1073.

Littell, R. C., Milliken, G. A., Stroup, W. W., and Wolfinger, R. D., 1996, *SAS System for Mixed Models,* SAS Institute, Inc., Cary, NC.

Salyani, M., 2000a, "Methodologies for assessment of spray deposition in orchard applications," *ASAE Paper No. 001031*, 11 pp.

Salyani, M., 2000b, "Optimization of deposition efficiency for airblast sprayers," *Transactions of the ASAE*, Vol. 43, No. 2, pp. 247-253.

Salyani, M. and Whitney, J. D., 1990, "Ground speed effect on spray deposition inside citrus trees," *Transactions of the ASAE*, Vol. 33, No. 2, pp. 361-366.

Salyani, M., Koo, Y. M., and Sweeb, R. D., 2000, " Spray Application Variables Affect Air Velocity and Deposition Characteristics of a Tower Sprayer," *Proceedings of the Florida State Horticultural Society*, Vol. 113: pp. (in press)

SAS Institute, Inc., 1990, *SAS User Guide: Statistics,* Ver. 6, SAS Institute, Inc., Cary, NC.

Whitney, J. D., 1999, "Field Test Results With Mechanical Harvesting Equipment in Florida Citrus," *Applied Engineering in Agriculture*, Vol. 15, No. 3, pp. 05-210.

Whitney, J. D., BenSalem, E., and Salyani, M., 2000, "The Effect of Trunk Shaker Patterns on Florida Orange Removal,"*ASAE Paper No.*, 001103, St. Joseph, MI., 9 pp.

Whitney, J. D., Hartmond, U., Kender, W. J., Burns, J. K., and Salyani, M., 1999, "Orange Removal with Trunk Shaker and Abscission Chemicals," *ASAE Paper No.*, 991078, St. Joseph, MI., 8 pp.

Wilcox, M. and Taylor, J. B., 1997, "Citrus Abscission at Very Low Application Rates," *Tropical Horticulture*, Vol. 41, pp. 45-151.

Wilson, W. C., Coppock, G. E., and Attaway, J. A., 1981, "Growth Regulators Facilitate Harvesting of Oranges," *Proceedings of the International Society of Citriculture*, pp. 278-281.

Milton E. Teske,[1] Andrew J. Hewitt,[2] and Robert E. Mickle[3]

The Measurement of Droplet Size Distributions from Rotary Atomizers

---

Reference: Teske, M. E., Hewitt, A. J., and Mickle, R. E., **"The Measurement of Droplet Size Distributions from Rotary Atomizers,"** *Pesticide Formulations and Applications Systems: A New Century for Agricultural Formulations, Twenty First Volume, ASTM STP 1414,* J. C. Mueninghoff, A. K. Viets, and R. A. Downer, Eds., American Society for Testing and Materials, West Conshohocken, PA, 2001.

Abstract: Various experimental techniques have been developed to recover droplet size distributions emitted from rotary atomizers. These techniques include both single line-of-sight measurements through the center of the spray cloud and volume weighted averaging from multiple passes through portions of the spray. Because the droplet size distribution of a spray is one of the most important inputs into aerial spray prediction models such as AgDRIFT®, a consistent, widely-accepted standard for rotary atomizers should be developed and implemented with regard to this measurement. No such standard currently exists. The present paper summarizes current measurement techniques and develops a simple computer model to predict droplet movement following emission from a rotary atomizer in a wind tunnel. Qualitative comparisons are made with available data to demonstrate the effects of single line-of-sight measurements and weighted averaging from multiple passes.

Keywords: Rotary atomizer, droplet size distribution, spray measurement

Introduction

The measurement of droplet size distributions for agricultural and biological sprays has been a subject of interest for many years. Different measurement systems such as laser diffraction, optical imaging, and Phase-Doppler light refraction can produce different apparent droplet size distributions for the same spray (Arnold 1990; Dodge 1987; Picot et al. 1993). These differences are due to several reasons, such as dynamic size range, sampling technique (number-density or number-flux weighted), data processing routines, and curve-fitting. Conversion routines have been proposed based on the first two issues (Teske et al. 2001b and Dodge 1992, respectively). Corrections for specific techniques can also be applied for sampling factors such as multiple scattering (Felton et al. 1985; Hovenac et al. 1992). A sampling issue that cannot so easily be corrected, however, is that of obtaining a representative cross-sectional average droplet size distribution through the spray cloud (Goguen et al. 1997). Most agricultural

---

[1] Senior associate, Continuum Dynamics, Inc., 34 Lexington Avenue, Ewing, NJ, USA, 08618.
[2] Specialist, Stewart Agricultural Research Services, Inc., P. O. Box 509, Macon, MO, USA, 63552.
[3] Consultant, REMSpC Spray Consulting, 12 Welsh Drive, RR3, Ayr, ON, Canada, N0B 1E0.

atomizers produce sprays with different droplet sizes at different locations within the spray cloud. For example, flat fan nozzles typically produce sprays where the droplets are larger at the edge of the fan than at its center. The accurate measurement of the droplet size distribution produced by such atomizers requires that a representative cross-sectional average sample be taken that includes all parts of the spray cloud. This effect can be achieved by sampling and averaging at multiple discrete positions throughout the spray cloud (Arnold 1983), or by continuously scanning the spray cloud across the laser beam (or vice-versa) while sampling (Dodge 1988; Snyder et al. 1989). It has been shown that samples taken only at single line-of-sight points within a spray cloud will not properly account for the variations in droplet size distribution throughout the spray. Combellack and Matthews (1981) found large differences in droplet sizes across spray clouds produced by the D2, D4, and D6 nozzle orifices with size 25 and 45 swirl plates, and different fan spray nozzles, as measured with a Malvern particle size analyzer. Chapple and Hall (1993) used a PDPA instrument to measure a detailed profile of the droplet size distribution produced by an 8003 nozzle, from which differences in the local particle size distribution were clearly visible. Lake and Dix (1985) used an X-Y nozzle transporter to attempt to obtain a representative sample with a PMS OAP instrument (other researchers have moved the PMS probes relative to the nozzle rather than moving the nozzle relative to the probe when sampling).

Hewitt et al. (1994b) described wind tunnel experiments spraying water with 0.25% v/v non-ionic surfactant at 2 bar pressure through D4-45 and 8004 nozzles mounted at 0 deg to a 50 m/s air stream. Results collected using a full vertical nozzle traverse were compared with averaged data collected from discrete sampling at 4, 7, and 11 equidistant positions throughout the spray clouds. The closest agreement in the cumulative droplet size distribution obtained using the full traverse occurred with the 4 and 7 sampling position data. The poorer agreement observed when using 11 sampling position data was attributed mainly to an increase in the spray volume contained in the larger droplet size fractions sampled at the periphery of the spray clouds. The outermost sample points also contained insufficient droplets to give a sample extinction above the Malvern recommended minimum level (0.01), and would therefore not be reliable. Hewitt et al. (1994b) stressed that the actual number and location of sampling positions used where a traverse is not possible will vary with the spray cloud width and variability.

When sampling droplet size distributions produced by rotary atomizers such as rotary cage, drum, and sleeve atomizers, and spinning discs, most workers have taken measurements at a single position (through the center of the hub), or a few positions (vertically below the hub), rather than taking a complete cross-sectional average as described above. This approach is followed because the spray cloud emitted at the periphery of rotary atomizers is approximately axisymmetric, rotating, and circular (if sampled with the atomizer front plate or disc straight back from the air stream), and does not produce the same three-dimensional spray structure associated with fan-spray atomizers. In effect the spray cloud may be considered hollow cone in shape, but with a much larger diameter than hollow cone nozzle spray clouds.

Hewitt (1997) conducted a wind tunnel study with a slotted rotary drum atomizer (described by Hewitt and Matthews 1991) to determine the differences in droplet size distribution measured with the atomizer positioned central in the wind tunnel, and traversed vertically through the working section, using the procedure developed for sampling hydraulic nozzle sprays in the Spray Drift Task Force (SDTF) atomization studies (Hewitt 1994). A slotted drum atomizer was selected in preference to a Micronair AU4000 or AU5000 rotary cage atomizer because the slotted drum was smaller (approximately 90 mm outer diameter and 50 mm drum length), allowing it to be fully traversed within the wind tunnel working section. In addition, it was powered by an electric motor, allowing the rotation rate to be precisely regulated independent of the air stream velocity. Blade-driven Micronair atomizers would be difficult to sample while traversing because the windmill blades or cage might contact the tunnel top and bottom during a traverse. The results of the study showed that droplet size distributions were very similar when sampling with and without a traverse. The study also showed that sampling with a traverse was very difficult, because spray coalescence on the wind tunnel walls when the atomizer was at the top of the tunnel could cause sampling errors. During these tests, the tunnel walls were covered with absorbent material, in an effort to reduce splash back and deter sampling of any splashed droplets.

The ASTM E29.04 Liquid Particle Size Measurement Subcommittee is developing a standard for measuring droplet size distributions using laser diffraction techniques. The draft standard, "*Standard Test Method for Determining Cross-Section Averaged Liquid Drop Size Characteristics in a Spray Using Laser-Diffraction Instruments,*" explains the importance of sampling for hydraulic nozzles but does not currently reference rotary atomizers where droplets are emitted centrifugally. Several researchers have reported distributions for rotary atomizers using several measurement techniques (Parkin et al. 1980; Spillman 1982; van Vliet and Picot 1987; Picot et al. 1989; Parkin and Siddiqui 1990; Hewitt and Matthews 1991; Skyler and Barry 1991; Hewitt et al. 1994a), not all consistent with one another. The need for extending this standard to include rotary atomizers will be explored here.

Modeling

Perhaps the best way to proceed is by illustration, developing a simple model for the behavior of droplets released from a rotary atomizer. For this exercise we assume that a Micronair AU4000 (with a hub diameter of 10 cm) spins at 8000 rpm, releasing spray material into a 50 m/s air stream, atomizing material into one of two droplet size distributions (Figure 1): an aerosol to very fine distribution (Matthews et al. 2001) and the ASAE S-572 very fine to fine distribution (Hewitt et al. 2001). The former is obtained from a green Sprays International Fogger Nozzle at 11 bar pressure, while the latter is generated by an 11001 nozzle at 3.8 bar pressure. These distributions represent the finer end of the droplet size distributions expected in aerial spraying, and are consistent with typical rotary atomizer spectra.

If we assume that the droplets all leave the Micronair AU4000 at the tangential speed of its rotary cage, we may write the appropriate Lagrangian equations of motion (Teske et al. 2001a) as

$$\frac{d^2 X_i}{dt^2} = [U_i - V_i] \left[\frac{1}{\tau_p} \right] + g_i \quad (1)$$

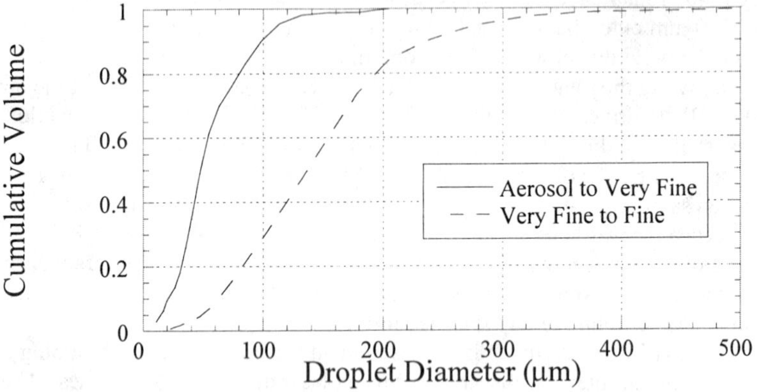

Figure 1 – *Aerosol to very fine ($D_{v0.5}$ = 47.8 μm) and ASAE S-572 very fine to fine ($D_{v0.5}$ = 137.2 μm) droplet size distributions.*

$$\frac{dX_i}{dt} = V_i \tag{2}$$

where X_i, V_i, and U_i are the ensemble-averaged ith components of droplet position, droplet velocity, and local fluid velocity, respectively, t is time, and g_i is gravity. The drag force on the droplet is represented by the relaxation time τ_p as

$$\tau_p = \frac{4}{3} \frac{D\rho}{C_D \rho_a |U_i - V_i|} \tag{3}$$

where D is the droplet diameter, ρ is the droplet density, ρ_a is the air density, and C_D is the drag coefficient, evaluated empirically for spherical droplets (Langmuir and Blodgett 1946) as

$$C_D = \frac{24}{Re}\left[1 + 0.197\, Re^{0.63} + 0.00026\, Re^{1.38}\right] \tag{4}$$

where the Reynolds number is defined by

$$Re = \frac{\rho_a D |U_i - V_i|}{\mu_a} \tag{5}$$

where μ_a is the viscosity of air. The relaxation time τ_p has physical significance with regard to dispersion, in that it is the e-folding time required for a droplet to catch up to its local velocity (for V_i to approach U_i). Equations 1 and 2 were first examined by Reed (1953) and, with a specification of the local velocity U_i, can be solved to obtain the trajectory paths for spray material issuing from the atomizer. In our application the wind tunnel air stream provides the local velocity.

While these equations include the effects of gravity, and provisions are made in the general model for evaporation, neither effect comes into play for the downstream distances considered here.

In general the rotary atomizer is mounted in a wind tunnel with its axis of rotation parallel to the air stream. Because the released droplets will continue along their radial trajectories until stopped by drag, it is more than likely that the wind tunnel walls will eventually collect the spray (at least for the larger droplets). Thus, there would appear to exist an optimal distance downstream of the atomizer at which to collect data with the desired measuring device. Three conditions seem to apply to this optimal distance: the distance must be close enough to the atomizer to (1) eliminate spray impaction on the lenses and (2) avoid the tunnel region where the walls, especially the ceiling, are impacted prior to sampling, but far enough downstream to (3) overcome number-density weighted (spatial) sampling effects, where a pronounced velocity profile within the spray is to be avoided (Hewitt and Valcore 1995). In our example we solve the above equations for snapshots of the droplet locations between 0.1 and 5.0 m downstream, as shown in Figure 2.

It may be seen from these figures that in the idealized problem the droplets form circular rings as they move outward from the hub. The smaller droplets stop in the shortest distances, staying close to the hub, while the larger droplets continue to move radially outward (concentric circles represent droplet size categories for the Malvern 800 mm lens). The drag effect may be seen by the darkened center of the pattern, containing the smaller droplet size categories. Droplet movement in the absence of gravitational effects would result in the radial distribution shown in Figure 3, clearly requiring a wind tunnel with a large cross-sectional area if the walls are to be avoided.

The simulation also includes the effect of setting the droplets into motion down the tunnel; these results then also recover the distances required for the droplets to reach (or nearly reach) the wind tunnel speed (Figure 4). At sampling distances of 0.5 to 1.0 m, only droplets with diameters less than 100 μm have approached the tunnel speed; larger droplets will therefore be sampled at speeds less than U_i. There is therefore an obvious compromise with regard to measurement distance downstream of the atomizer, because the droplets should achieve tunnel speed and yet not hit the walls before being counted.

What is also of particular interest – and what is most easily seen in Figure 2 – is that the recovery of droplet size distribution data anywhere below the centerline of the hub will always result in a miss of the smaller droplet sizes. Such a miss may have dire consequences if the intent of the wind tunnel study is to recover the most representative droplet size distribution as input into a drift prediction model such as AgDRIFT®. For example, if a pass were made 10 cm below the atomizer hub at a distance of 0.5 m downstream, Figure 2 shows that approximately 40% of the droplet categories will be missed by the measurement. These droplet categories contain the smaller droplets in the droplet size distribution and have a stronger tendency to drift than the larger droplets in

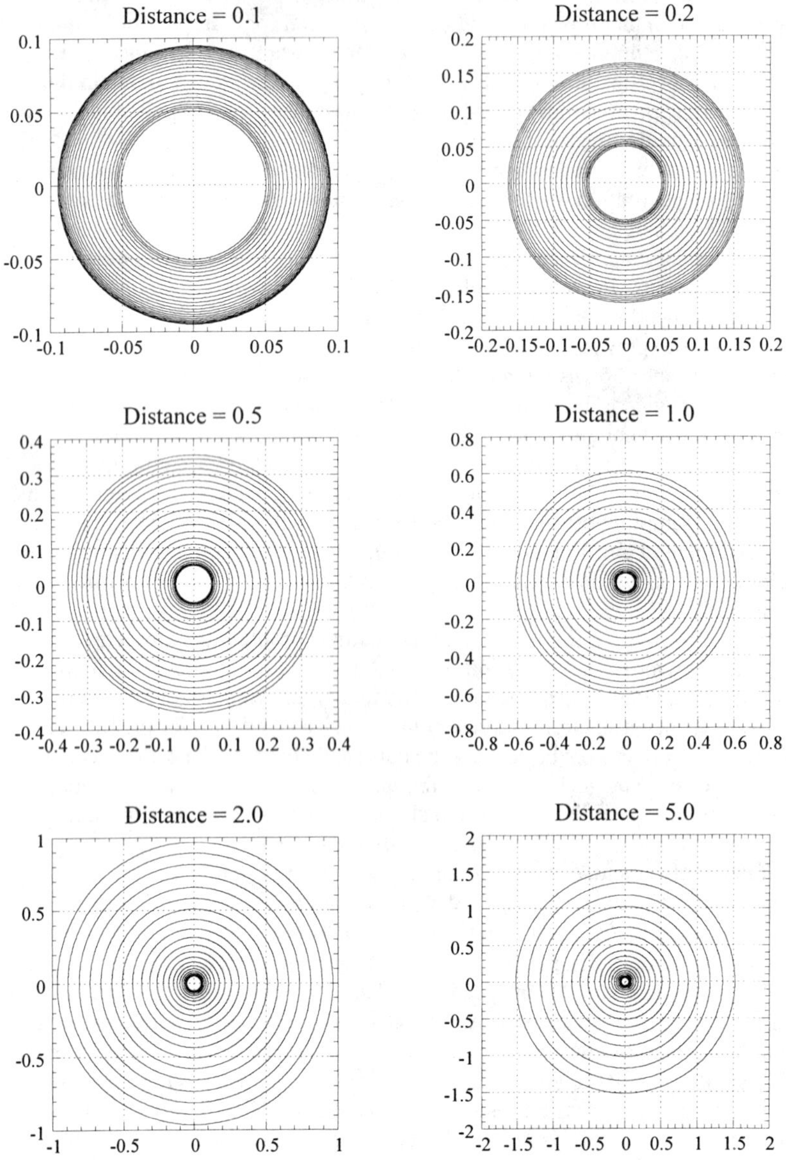

Figure 2 – *Rotary atomizer downstream droplet patterns. The horizontal and vertical axes in each snapshot span the width and height of the droplet pattern at the downstream distances shown, one circle for each droplet size category (all distances are in m).*

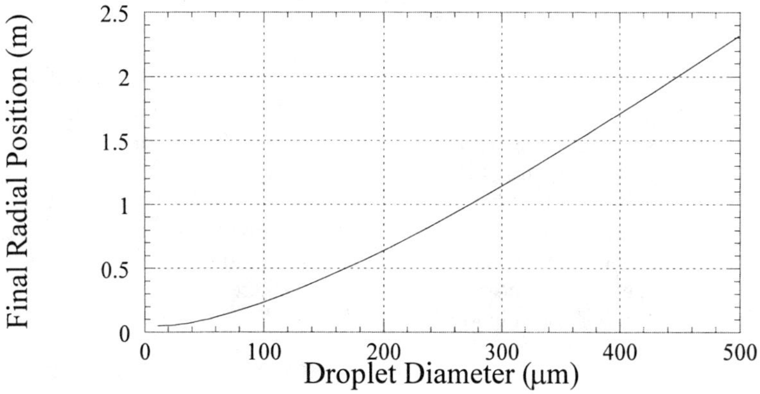

Figure 3 – *Final droplet radial positions in the absence of gravitational effects.*

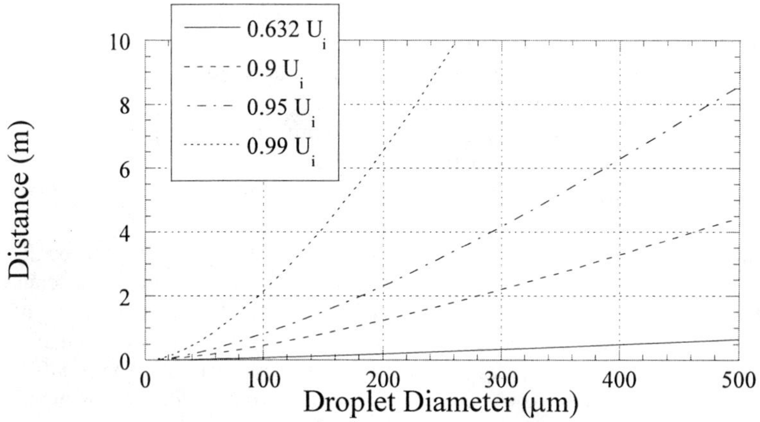

Figure 4 – *Downstream distances required for droplets to approach the wind tunnel speed. The notation identifying each curve indicates the downstream speed achieved by any size droplet as a fraction of the wind tunnel speed.*

the distribution. Missing the presence of the smaller droplets in the droplet size distribution could lead to an under-prediction of drift from rotary atomizers.

This effect may also be seen in Figure 5, where we compare the results of a single line-of-sight measurement through the center of the hub (replicating the droplet size distributions shown in Figure 1), with results generated by volume averaging multiple passes through the spray cloud at several distances vertically from the hub. The volume averaging was achieved by moving the instrument further and further vertically from the atomizer hub and progressively missing one droplet category with each measurement. These multiple passes bias the resulting droplet size distributions toward their larger droplet categories, and under-predict the number of finer droplets most responsible for drift. We would expect the actual averaging process from multiple scans to recover droplet size distributions that lie between the limits shown here.

The effect on $D_{v0.5}$ may be seen in Figure 6 for the two droplet size distributions considered here.

Data Comparison

To test the appropriateness of our model predictions, we compare our results to data (Hewitt 1997). Figure 7 presents the cumulative volume recovered at different chordal positions for a water spray through an AU5000 at 6000 rpm in a 50 m/s air stream. The Malvern instrument was positioned 15 cm downstream from the hub of the rotary atomizer. Droplet size distribution measurements were taken through the center of the spray cloud and at five chordal positions located progressively further vertically from the center of the cloud. The chordal measurements were averaged by droplet size class volume. Note the decrease in cumulative volume in the smaller droplet sizes as the chordal position increases (from left to right on the figure), and the general increase in droplet sizes with the volume weighted average over and above the spectrum recorded at the hub. In this example $D_{v0.5}$ for the centerline spectrum is 70.4 μm, while for the volume weighted average it is 111.9 μm.

A second series of Malvern measurements – this time for Foray 48B neat through an AU4000 at 7500 rpm in a 50 m/s air stream – is shown in Figure 8. Here the differences in the several chordal measurements (again taken 7.5 cm downstream from the hub of the rotary atomizer) are less obvious than in Figure 7, and suggest that finer sprays may not be as susceptible to characterization difficulties. The windmill blades on the AU4000 extend from the hub and effectively occupy a substantial portion of the cross sectional area of the wind tunnel. Splash back from the tunnel walls, loss of mean wind tunnel speed as the walls are approached by the droplets, and finer spray quality may account for some of the differences shown here, particularly for the chordal positions farthest from the hub. In this example $D_{v0.5}$ for the centerline spectrum is 66.5 μm, and varies between 62.6 μm and 75.1 μm for the chordal measurements.

These plots compare qualitatively with the model results shown in Figure 5 and demonstrate the effect of averaging chordal data. Our model appears to have captured the essence of the measurement problem hypothesized in this paper.

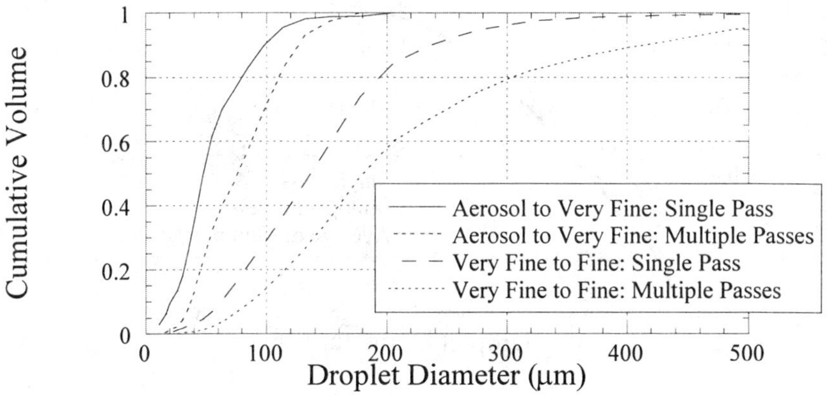

Figure 5 – *Cumulative volume results for multiple passes through the spray cloud. The single pass is through the center of the hub of the rotary atomizer, while multiple passes traverse several vertical distances above or below the hub.*

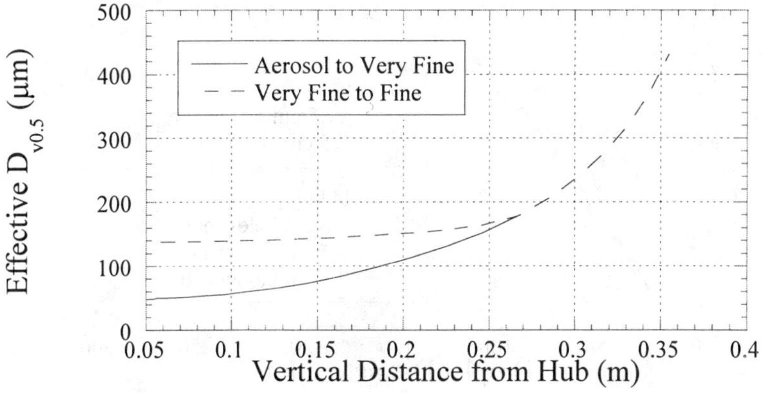

Figure 6 – *Effects on $D_{v0.5}$ of scanning the spray cloud further and further vertically from the hub, in this case at a downstream distance of 0.5 m.*

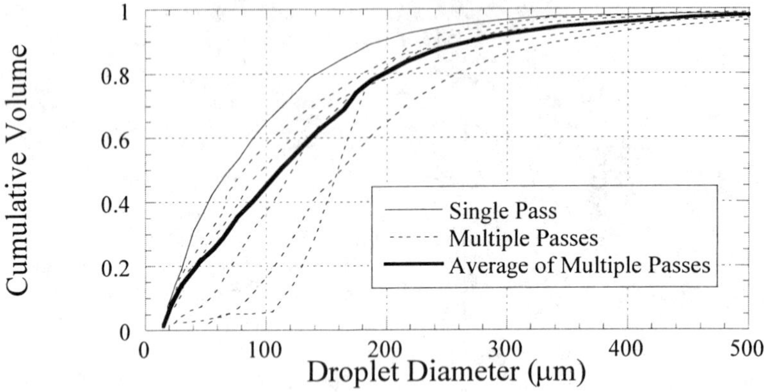

Figure 7 – *Data comparison between single and multiple pass droplet size distributions in the wake of a Micronair AU5000, and the average of the chordal data. The single pass is traversed across the center of the hub of the rotary atomizer, while the multiple passes are made at different vertical distances above or below the hub center.*

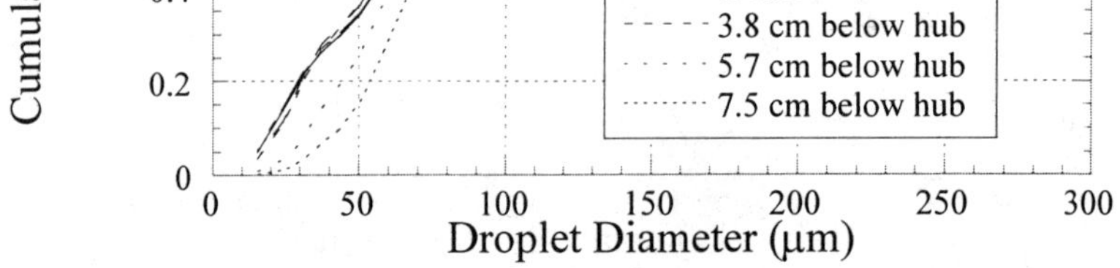

Figure 8 – *Data comparison for several passes in the wake of a Micronair AU4000.*

Conclusions

This research has shown that a need exists for the standard measurement of droplet size distributions from rotary atomizers. Current techniques that include averaging chordal measurements through the spray cloud may overemphasize the larger droplets and hence result in a droplet size distribution with a larger $D_{v0.5}$ than actually exists.

The proposed standard should address the following three issues: (1) determining the location of the instrument with respect to the rotary atomizer in order to minimize the sampling error due to spatial-temporal effects; (2) establishing the measurement method to follow (a single line-of-sight measurement through the center of the spray cloud or a weighted average from multiple passes through portions of the spray); and (3) including mass balance as a means of ensuring a complete scan of the spray cloud.

Such a standard has been prepared and was being revised following balloting through ASTM at the time of preparation of this paper ("*Standard Test Method for Determining Cross-Section Averaged Liquid Drop Size Characteristics in a Spray Using Laser-Diffraction Instruments*"). Sampling for rotary atomizer sprays is covered in the standard, with recommendations on optimal sampling being based on the findings of the present research. The draft standard included the following statement in this respect: "Teske et al. (2002) [referencing this paper] conducted research suggesting that the sprays produced by rotary atomizers may not need to be sampled using multiple chordal locations if the atomizer is mounted in a co-flowing air stream in such a way to ensure that droplets are carried across the laser beam. For such sampling, a single centerline measurement is sufficient to provide droplet size spectrum information that includes the smaller droplets within the spray that may otherwise be under-estimated with measurement locations above or below the centerline."

References

Arnold, A. C., 1983, "Comparative Droplet Size Spectra for Three Different Angled Flat Fan Nozzles," *Crop Protection*, Vol. 2, No. 2, pp. 193-207.

Arnold, A. C., 1990, "A Comparative Study of Drop Sizing Equipment for Agricultural Fan-Spray Atomizers," *Aeronautical Science and Technology*, Vol. 12, pp. 431-445.

Chapple, A. C. and Hall, F. R., 1993, "A Description of the Droplet Spectra Produced by a Flat-Fan Nozzle," *Atomization and Sprays*, Vol. 3, No. 4, pp. 477-88.

Combellack, J. H. and Matthews, G. A., 1981, "Droplet Spectra Measurements of Fan and Cone Atomizers Using a Laser Diffraction Technique," *Journal of Aerosol Sciences*, Vol. 12, No. 6, pp. 529-540.

Dodge, L. G., 1987, "Comparison of Performance of Drop-Sizing Instruments," *Applied Optics*, Vol. 27, pp. 1328-1341.

Dodge, L. G., 1988, "Representation of Average Drop Sizes in Sprays," *Journal of Propulsion & Power*, Vol. 4, No. 6, pp. 490-496.

Dodge, L. G., 1992, "TESS: Tool for Spray Studies," *Technology Today*.

Felton, P. G., Hamidi, A. A., and Aigal, A. K., 1985, "Measurement of Drop Size Distribution in Dense Sprays by Laser Diffraction," Proceedings of ICLASS Third International Conference, London, England, pp. 1-13.

Goguen, R. J., Saunders, B. A. J., and Picot, J. J. C., 1997, "Flux-Based Measurements with the Malvern Spectrometer," *Transactions of the ASAE*, Vol. 40, No. 5, pp. 1233-1236.

Hewitt, A. J. and Matthews, G. A., 1991, "Some Aspects of Rotary Atomizer Performance," Proceedings of ILASS-Europe, Swansea, UK, pp. 200-206.

Hewitt, A. J., 1994, "Determining Cross-Section Average Drop Size Distributions," Standard Operating Procedure SDTF SOP 11A/5, Spray Drift Task Force.

Hewitt, A. J., Robinson, A. G., Sanderson, R., and Huddleston, E. W., 1994a, "Comparison of the Droplet Size Spectra Produced by Rotary Atomizers and Hydraulic Nozzles Under Simulated Aerial Application Conditions," *Journal of Environmental Science & Health*, Part B, Vol. 29, No. 4, pp. 647-660.

Hewitt, A. J., Sanderson, R., Huddleston, E. W., and Ross, J. B., 1994b, "Approaches to Measuring Atomization Droplet Size Spectra Using Laser Diffraction Particle Size Analyzers in Wind Tunnels," Proceedings of ILASS-Americas Seventh Annual Conference, Bellevue, WA, pp. 116-120.

Hewitt, A. J. and Valcore, D. L., 1995, "Measurement Techniques for Simulated Agricultural Sprays Produced by Ground Sprayers Using Number Density Weighted Sampling Techniques," Proceedings of ILASS-Americas Eighth Annual Conference, Troy, MI, pp. 29-33.

Hewitt, A. J., 1997, "Spray Drift Task Force Miscellaneous Nozzle Study," SDTF Report No. A95-010, EPA MRID No. 44310401.

Hewitt, A. J., Johnson, D. R., Fish, J. D., Hermansky, C. G., and Valcore, D. L., 2001, "The Development of the Spray Drift Task Force Database on Pesticide Movements for Aerial Agricultural Spray Applications," *Environmental Toxicology and Chemistry* (to appear).

Hovenac, E. A., Oldenburg, J. R., and Lock, J. A., 1992, "Experimental Testing of Four Correction Algorithms for the Forward Scattering Spectrometer Probe," NASA Technical Report No. 105906.

Lake, J. R. and Dix, A. J., 1985, "Measurement of Droplet Size with a PMS Optical Array Probe Using an X-Y Nozzle Transporter," *Crop Protection*, Vol. 4, No. 4, pp. 464-472.

Langmuir, I. and Blodgett, K. B., 1946, "A Mathematical Investigation of Water Droplet Trajectories," Army Air Force Technical Report No. 5418.

Matthews, G. A., Bateman, R., and Hewitt, A. J., 2001, "Classification Criteria for Fog and Mist Application of Pesticides," *Atomization and Sprays* (in review).

Parkin, C. S., Wyatt, J. and Wanner, R., 1980, "The Measurement of Drop Spectra in Agricultural Sprays Using a Particle Measuring Systems Optical Array Spectrometer," *British Crop Protection Council Monograph 24*, pp. 241-249.

Parkin, C. S. and Siddiqui, H. A., 1990, "Measurement of Drop Spectra from Rotary Cage Aerial Atomizers," *Crop Protection*, Vol. 9, pp. 33-38.

Picot, J. J. C., van Vliet, M. W., and Payne, N. J., 1989, "Droplet Size Characteristics for Insecticide and Herbicide Spray Atomizers," *Canadian Journal of Chemical Engineering*, Vol. 67, pp. 752-761.

Picot, J. J. C., Goguen, R., and Murray, D., 1993, "Wind Tunnel Atomization Characterizations with the Malvern Spectrometer: The Spatial Sampling Error," Proceedings of ILASS-Americas Sixth Annual Conference, Worcester, MA, pp. 157-161.

Reed, W. H., 1953, "An Analytical Study of the Effect of Airplane Wake on the Lateral Dispersion of Aerial Sprays," NACA Report No. 3032.

Skyler, P. J. and Barry, J. W., 1991, "Final Report - Compendium of Drop Size Spectra Compiled from Wind Tunnel Tests," USDA Forest Service Report No. FPM 90-9.

Snyder, H. E., Senser, D. W., and Lefebvre, A. H., 1989, "Experimental Techniques for Drop-Size Measurements in Fan Sprays," Proceedings of ILASS-Americas Sixth Annual Conference, Worcester, MA, pp. 56-61.

Spillman, J., 1982, "Atomizers for the Aerial Application of Herbicides – Ideal and Available," *Crop Protection*, Vol. 1, No. 4, pp. 473-482.

Teske, M. E., Bird, S. L., Esterly, D. M., Curbishley, T. B., Ray, S. L., and Perry, S. G., 2001a, "AgDRIFT®: A Model for Estimating Near-Field Spray Drift from Aerial Applications," *Environmental Toxicology and Chemistry* (to appear).

Teske, M. E., Thistle, H. W., Hewitt, A. J., and Kirk, I. W., 2001b, "Conversion of Droplet Size Distributions from PMS Optical Array Probe to Malvern Laser Diffraction," *Atomization and Sprays* (to appear).

van Vliet, M. W. and Picot, J. J. C., 1987, "Drop Spectrum Characterization for the Micronair AU4000 Aerial Spray Atomizer," *Atomization and Spray Technology*, Vol. 3, pp. 123-134.

SESSION VII: Formulation Ingredients II

Johan C. G. Rommens[1] and Simon J. Davies[2]

Monobranched Alcohol Alkoxylates: A New Generation of Surfactants for Use in Crop Protection

---

Reference: Rommens, J. C. G. and Davies, S. J., "**Monobranched Alcohol Alkoxylates: a New Generation of Surfactants for Use in Crop Protection**", *Pesticide Formulations and Application Systems: A New Century for Agricultural Formulations, Twenty First Volume, ASTM STP 1414*, J. C. Mueninghoff, A. K. Viets, and R. A. Downer, Eds., American Society for Testing and Materials, West Conshohocken, PA, 2001.

Abstract: Monobranched alcohol alkoxylates (Atplus® MBA and Atlox® MBA) are a special class of alcohol alkoxylates with specific properties related to the monobranched structure of the hydrophobe. The availability of different alkyl chain lengths (C11 to C15) and ethoxylation / alkoxylation degrees (from 6 to 30 EO) results in a range of surfactants which has been selected as preferred replacement products for nonyl phenol ethoxylates without carrying a hazard classification. The particular type of branching of the hydrophobe results in lower gelling and foaming tendency compared to linear alcohol alkoxylates with similar carbon chain lengths and alkoxylation degree. The compact nature of the hydrophobe is responsible for the very fast dynamic behaviour which makes MBAs very suitable as adjuvants for pesticides. Furthermore, MBAs show lamellar liquid crystalline phases which are less viscous at higher concentrations than linear counterparts. Monobranched alcohol alkoxylates have shown to be efficient in enhancing the bio-efficacy of selective and non-selective herbicides and fungicides.

Keywords: monobranched alcohol alkoxylate, NPE replacement, adjuvant

Introduction

Because of regulatory pressure in crop protection the use of co-formulants has become necessary to enhance the efficacy of agrochemically active ingredients. Alcohol alkoxylates are widely commercially available as non-ionic surfactants and are applied as wetters, adjuvants and emulsifiers in crop protection. However, not only the efficacy of the surfactant but also the environmental profile has gained importance. Nonyl phenol ethoxylates (NPEs) were amongst the first commercially available nonionic surfactants. However the recent concerns related to the potential endocrine modulating properties of NPEs have stressed once more the importance of an acceptable environmental profile.

---

[1] Technical manager, Uniqema, 45 Everslaan, B-3078 Everberg, Belgium
[2] Senior research scientist, ICI, PO Box 90 Wilton Centre, Middlesbrough TS90 8JE, UK

Monobranched alcohol (MBA) alkoxylates are a special class of alcohol alkoxylates with specific properties related to themonobranched structure of the hydrophobe. The availability of different alkyl chain lengths (C11 to C15) and ethoxylation degrees (from 6 to 30 EO) results in a range of surfactants, which has been selected as preferred replacement products for nonyl phenol ethoxylates without carrying a hazard classification and readily biodegradable (Reekmans 1998). A range of MBA ethoxylates was selected that is exempted from the requirement of an EPA tolerance for use in pesticide formulations under 40 CFR 180.1001 (c) and (e) (Table 1). Propoxylation of the MBA ethoxylates resulted in a range of MBA alkoxylates exempted from the requirement of an EPA tolerance for use in pesticide formulations under 40 CFR 180.1001 (d).

Table 1 - *Overview and chemical composition of the surfactants used in this study.*

Product[3]	Description [(a)]
Atlox® MBA 11/6	C11 MBA POE(6)
Atlox® MBA 11/10	C11 MBA POE(10)
Atlox® MBA 13/7	C13 MBA POE(7)
Atlox® MBA 13/15	C13 MBA POE(15)
Atplus® MBA 1301	C13 MBA POE / POP
Atplus® MBA 1303	C13 MBA POE / POP
Atlox® MBA 1306	C13 MBA POE / POP (high MW)
Atlas® G 5000	polyalkylene glycol ether
Atlox® 4838B	calcium dodecylbenzene sulfonate in alcohol
Synperonic® NP8	nonyl phenol POE (8)
Lubrol® 17A17	oleyl alcohol POE (17)
Synperonic® L7	lauryl alcohol POE (7)
Synperonic® PE/L64	alkoxylated block copolymer
Synperonic® A7	C13/15 alcohol POE (7)

[(a)]*POE : Polyethylene oxide ; POP : Polypropylene oxide*

MBA Alkoxylates as Formulation Additives

The particular type of branching of the hydrophobe results in lower gelling and foaming tendency compared to linear alcohol alkoxylates with similar carbon chain lengths and alkoxylation degree.

Furthermore, MBAs show much lower viscosity liquid crystal phases at high surfactant concentrations than linear counterparts. This is because of the nature of the liquid crystal phases formed. The bulkier hydrophobe of the branched alcohol

[3] Atlox® MBA 11/6, Atlox® MBA 11/10, Atlox® MBA 13/7, Atlox® MBA 13/15, Atlox® MBA 1306, Atplus® MBA 1301, Atplus® MBA 1303, Atlas® G 5000, Atlox® 4838B, Synperonic® NP8, Lubrol® 17A17, Synperonic® L7, Synperonic® PE/L64, Synperonic® A7, ICI Belgium n.v., B-3078 Everberg, Belgium

ethoxylates makes it much more difficult for these surfactants to form highly viscous hexagonal and cubic liquid crystal phases. Rather lamellar phases alone are formed (Table 2). This difference in viscosity behaviour will be important in the ease of use of the surfactant system.

Table 2 - *Phase behaviour of MBA ethoxylates compared to other alcohol ethoxylates for different surfactant concentrations*[a].

Chemical Description	Concentration (% w/w)							
	20	30	40	50	60	70	80	90
C13 MBA POE(7)	L_a	L_a	L_a	L_a	L_a	L_a	L_a	L_2
lauryl alcohol POE(7)	L_1	L_1	H_1	H_1	V_1/L_a	L_a	L_a	L_2

[a] *The phase behaviour was examined under the optical microscope using polarised light. L_1 =isotropic micellar phase; L_a = lamellar phase; V_1 =cubic phase; H_1 =hexagonal phase; L_2 =inverse micelle phase*

Use in Emulsifiable Concentrates (EC)

Historically, it has been found that greatest versatility within an emulsifier system is accomplished by considering blends of:

1. A water insoluble anionic surfactant (for example, calcium dodecylbenzene sulfonate);
2. A low water-soluble ethoxylated nonionic surfactant (for example, nonyl phenol POE (6));
3. A high water-soluble ethoxylated nonionic surfactant (for example, nonyl phenol POE (30)); and
4. A high molecular weight water-soluble nonionic surfactant (for example, EO/PO block co-polymers)

As an alternative class to NPEs, MBA alkoxylates were evaluated. MBA 11/6 and MBA 11/10 were evaluated as substitutes for the low water-soluble and high water-soluble ethoxylates above, respectively. In addition, a monobranched alcohol POE-POP alkoxylate, was substituted for the high molecular weight water soluble nonionic surfactant. Use of an alternate high molecular weight polyalkylene glycol ether is also possible and will be equally effective. As anionic surfactant a solution of calcium dodecylbenzene sulfonate in alcohol was used (CaDBS).

Using an experimental design approach the ability of this emulsifier system to emulsify several actives was evaluated. In these experiments the concentration of the active ingredient was kept constant at 42% (w/w) and the total emulsifier concentration was 8.5 % (w/w). An aromatic solvent[4] was used. Two experiments were carried out. In the first experiment (Figure 1a) the concentration of the CaDBS solution was kept constant at 4.25 % (w/w) and the concentrations of the three MBA surfactants were varied within the remaining 4.25 % (w/w). In the second experiment (Figure 1b) the concentration of the CaDBS solution was kept constant at 5.1 % (w/w) and the concentrations of the three MBA surfactants were varied within the remaining 3.4 % (w/w).

---

[4] Aromatic® 150, ExxonMobil Chemical Belgium, B-2030 Antwerpen, Belgium

The composition of stable ECs based on these three MBA surfactants can be visualised on a triangle (Figure 1).

The apices of the triangle (e.g. Figure 1a) represent the formulations with 4.25 % of one MBA surfactant (e.g. the top of the triangle : Nr 1 represents a formulation with 4.25 % MBA 11/10 and no MBA 1306 or MBA 11/6).

The perimeter of the triangle represents the ECs with only two MBA surfactants (e.g. Nr 16 : 2.125 % MBA 11/10 and 2.125 % MBA 1306).

The area of the triangle represents a three component blend of MBA surfactants.

All formulations were evaluated on storage stability over 24 hours at room temperature and upon dilution (5%) in 342 ppm water. Only formulations which did not show phase separation after 24 hours storage (undiluted and upon 5 % dilution) are indicated on the triangle in Figure 1 as stable ECs. Following this approach stable ECs with permethrin, malathion, metalochlor, acetochlor and diazinon were successfully developed.

(a)

MBA 11/10

1
2 3
4 5 6
7 8 9 10
11 12 13 14 15
16 17 18 19 20 21
22 23 24 25 26 ***d*** 28
29 30 31 32 34 35 ***d*** 37
38 39 40 41 42 ***d*** ***d*** 45 46
47 48 49 50 51 ***p*** ***p*** ***a*** 55 56
57 58 59 60 61 62 ***p*** ***a*** ***a*** 66 67

MBA 1306 ***MBA 11/6***

(b)

MBA 11/10

1
d 3
a 5 6
me ***me*** 9 10
a/me ***ma/me*** 13 14 15
me/d ***ma/d*** ***ma*** 8 19 20 21
d/ma ***d/me*** ***a*** ***a*** 26 27 28
ma/d ***a/ma*** ***me/ma*** ***a*** 34 35 36 37
ma ***a/d*** ***ma/d*** ***a/p*** ***p*** 43 44 45 46
ma ***me/ma*** ***a/ma*** ***me/a*** ***p*** ***p*** 53 54 55 56
ma ***ma*** ***a/me*** ***a*** ***p*** ***p*** ***p*** 64 65 66 67

MBA 1306 ***MBA 11/6***

Figure 1 : *Composition of ECs based on MBA emulsifiers in combination with CaDBS solution. Concentration of the active ingredient was kept constant at 42% (w/w) and the total emulsifier concentration was 8.5 % (w/w). The letters on the triangles are referring to the compositions of stable formulations with the following actives : p : permethrin ; ma : malathion ; me : metalochlor ; a : acetochlor ; d : diazinon (a) 4.25 % CaDBS solution (i.e. 4.25 % MBA surfactants) (b) 5.1 % CaDBS solution (i.e. 3.4 % MBA surfactants).*

Use in Concentrated Emulsions (EW)

Using MBA 11/6 as emulsifier, an aromatic solvent[5] could be emulsified in water at 80%W/W. An alkoxylated block polymer was added to prevent oil droplet agglomeration upon storage of the formulation. The emulsion (Table 3) did not show any phase separation after 24 hours storage at room temperature. Also 5 % dilution of the

[5] Aromatic® 200, ExxonMobil Chemical Belgium, B-2030 Antwerpen, Belgium

emulsion in 324 ppm water did not show any phase separation after 24 hours storage at room temperature.

Table 3 - *Composition of an EW with MBA 11/6 as emulsifier.*

Formula	% W/W
aromatic solvent	80
C11 MBA POE(6)	6
alkoxylated block copolymer	2
water	12

MBA Alkoxylates as Adjuvants in Crop Protection

If one considers how adjuvants influence the biological efficiency of pesticides, the most logical approach is to evaluate how adjuvants affect the individual processes occurring during and after the spray application of the pesticide formulation (Holloway 1998, Tadros 1994).

Retention

An important process is droplet adhesion and retention. The low surface tension / contact angle predicts already excellent wetting and spreading properties for the MBA ethoxylates (Table 4).

Table 4 - *Physico-chemical parameters of alcohol ethoxylates.*

Chemical description	cmc[(1)]	γ_e[(2)]	SF[(3)]	τ_D[(4)]	ε_0[(5)]
C11 MBA POE(7)	3.10^{-4}	27	3.3	7	65
C13 MBA POE(15)	4.10^{-5}	32	1.6	32	50
C13/15 synthetic alcohol POE(7)	9.10^{-6}	29	-	180	68
oleyl alcohol POE(17)	7.10^{-7}	38	1.5	1200	45
nonyl phenol POE (8)	7.10^{-5}	29	2.1	108	70

[(1)]*Critical micellar concentration, mol/l*

[(2)]*Equilibrium surface tension (EST) of a 0.1% (w/w) solution measured at 20°C, mN/m*

[(3)]*Spread factor of a 0.2% (w/w) surfactant solution on pea, calculated as the ratio of the diameter of the wetted area and the diameter of the applied drop.*

[(4)]*Decay time constant from dynamic surface tension experiments, ms (surfactants at 0.2%w/w)*

[(5)]*Instantaneous dilational (elastic) modulus, mN/m (measured at cmc)*

However, for surfactant-containing systems there is no correlation between the equilibrium surface tension of the system and spray retention. This is because during the impaction process the surface of the droplet is rapidly expanded and the unmicellised surfactant cannot adsorb quickly enough at the newly created surface to maintain the equilibrium surface tension. A model was developed that correlates an "effective surface tension" directly to the spray retention. This effective surface tension is calculated based on the rate of surfactant adsorption at an interface and the instantaneous dilational modulus of the surfactant containing surface (Reckmans 1998).

The rate of surfactant adsorption, when a surfactant is present at a concentration above its cmc, is primarily determined by the value of the cmc. The higher the value of the cmc and the shorter the lifetime of the micelles, the higher the concentration of unmicellised surfactants and the faster the surfactant adsorption. The compact nature of the hydrophobe in the MBA alkoxylates results in higher cmc values and hence a fast adsorption rate. End capping of the MBA ethoxylates with a propylene oxide unit will result in smaller micelles with reduced stability. This also results in an increased adsorption rate.

The instantaneous dilational modulus depends upon the packing of the surfactant molecules at the interface with the lower the amount of surfactants per unit surface area the lower the value of the modulus (if the surfactant is behaving ideally). End capping of the ethylene oxide groups with propylene oxide will also increase the head group size and result in a reduced dilational modulus and hence an improved retention.

In Figure 2 the correlation between this effective surface tension and the retention at pea leaves is shown. It can be observed that MBA 11/6 and MBA 1303 (POP end capped) give an excellent retention. The effect of end capping is seen when MBA 1303 is compared with MBA 13/7 (no POP).

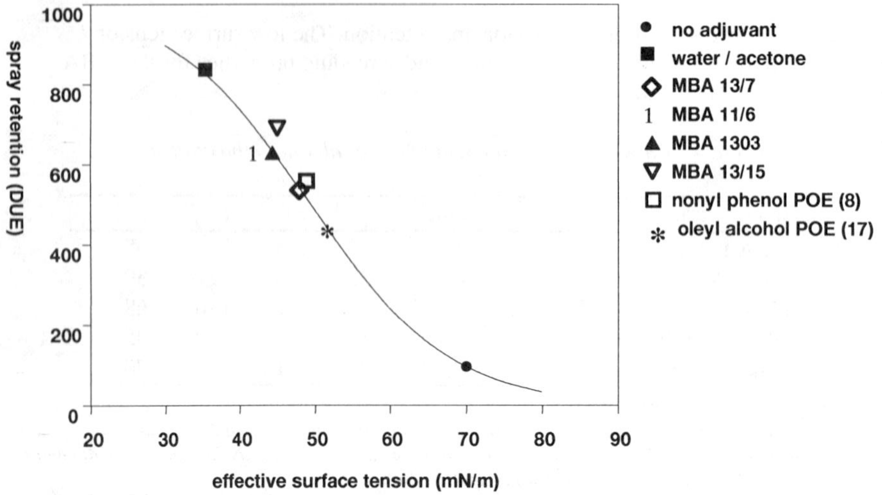

Figure 2 - *Spray retention on pea against effective surface tension (surfactant conc. = 1g/l in water). Fluoresceine was included in the formulation and retention determined via fluorescence measurements. DUE : data expressed as deposit per unit emission.*

Uptake

Next to wetting, spreading and spray retention, also uptake of the active ingredient into the leaf surface can be enhanced by using adjuvants. As already mentioned above, MBA alkoxylates show much lower viscosity liquid crystal phases at high surfactant concentrations than linear counterparts. This could be critical in the speed of uptake of the active component, with the active diffusing much more rapidly through the lower viscosity deposits (Table 5).

Table 5 - *% Uptake of radiolabeled haloxyfop-methyl into oat leaves.*

Adjuvant(1)	% uptake(2) after 5 h	% uptake after 24 h
MBA 1301	42.7	88.5
oleyl alcohol POE(17)	24.8	77.2
no adjuvant	27.8	69.7

(1)The adjuvant was applied at 1g/l. The herbicide concentration was 0.26 g/l (Half of the Normal Application Rate (HNAR)).
(2)Cellulose acetate film-stripping followed by liquid scintillation counting (LSC) was used to monitor the amounts of radioactivity remaining on the surface of treated leaf areas (Silcox and Holloway 1986). Recovery values were converted to percentage uptake with reference to the radioactive doses originally applied.

Field trials

Several field trials have confirmed that MBA alkoxylates are excellent adjuvants for herbicides and fungicides as can be seen in the two examples below (Tables 6 and 7).

Table 6 - *Effect of MBA 1301 on the efficacy of different concentrations haloxyfop-R-methyl on Italian ryegrass*

DAT(1)	Adjuvant(2)	haloxyfop-R-methyl (g/ha)(3)			
		54	37	18	9
7	no adjuvant(4)	59	55	32.5	0
7	MBA 1301	67.5	62.5	40	6
13	no adjuvant(4)	82.5	77.5	30	2
13	MBA 1301	96	83	60	5
25	no adjuvant(4)	98	92	0	0
25	MBA 1301	100	94.5	52.5	0

The formulation was sprayed at 200 l/ha.
(1) Evaluations were done 7, 13 and 25 days after treatment (DAT)
(2) The adjuvants were applied as tank mix at 125 g/ha.
(3) haloxyfop-R-methyl was applied at several doses : 54, 37, 18 and 9 g/ha. Evaluation was expressed as % growth reduction relative to the nearest untreated control plot (=0%).
(4) Gallant® formulation[6]

Table 7 : *Effect of MBA ethoxylates on the efficacy of propiconazole against winter wheat infected with Septoria tritici.*

Application	% infected leaf area(3) (2 top leaves)	% infected leaf area(3) (3 top leaves)
Untreated	50	50

[6] Gallant® , Dow Agroscicnes, Indianapolis, IN 46268-1054, USA

Tilt NAR(1)	5	25
Tilt 75 % NAR(1)	10	35
MBA 11/8(2)	5	20
MBA 11/7(2)	2.5	25

(1)Propiconazole (250 g/l)[7] was sprayed at 0.500 l/ha (Normal Application Rate (NAR)) and 0.375 L/ha (75 % NAR).

(2)MBA ethoxylates were added at 0.1 % (tank mix) to Tilt at 75 % NAR.

(3)The evaluation was done as % infected leaf area, 3 weeks after treatment.

Conclusion

In this paper the versatile use of monobranched alocohol alkoxylates in crop protection has been described. The favourable environmental profile, together with excellent properties as emulsifier in emulsifiable concentrates and concentrated emulsions turns this range into valuable replacement products for NPEs. The low viscosity phase behaviour does not only improve the physical form of MBA alkoxylates, compared to linear countertypes, but also results in beneficial properties as adjuvant. Thanks to the study of structure activity relationships it is better understood why this range of surfactants is enhancing the efficacy of many crop protection products.

References

Holloway, P. J., 1998, “Improving agrochemical performance : possible mechanisms for adjuvancy,” *Chemistry and Technology of Agrochemical Formulations*, Ed. D. A. Knowles, Kluwer Academic Publishers, Dordrecht, Boston, London, pp 232-260

Reekmans, S., 1998, “Novel surfactants and adjuvants for agrochemicals,” *Chemistry and Technology of Agrochemical Formulations*, Ed. D. A. Knowles, Kluwer Academic Publishers, Dordrecht, Boston, London, pp 179-226

Silcox & Holloway, 1986, *Aspects of Applied Biology*, 11, pp 13-17

Tadros, Th. F., 1994, *Surfactants in Agrochemicals*, Ed. Marcel Dekker Inc., New York, pp 207-248

[7] Tilt® EC, Syngenta, CH-4002 Basel, Switzerland

Frank J. Lachut [1]

Foam Control Strategies for Formulating with Alkypolyglycoside Surfactants

---

Reference: Lachut, F.J., **"Foam control Strategies for Formulating with Alkylpolyglycoside Surfactants,"** *Pesticide Formulations and Application Systems: A New Century for Agricultural Formulations, Twenty First Volume, ASTM STP 1414,* J. C. Mueninghoff, A. K. Viets, and R. A. Downer, Eds., American Society for Testing and Materials, West Conshohocken, PA, 2001.

Abstract: Alkylpolyglycosides are versatile, mild and environmentally friendly surfactants derived from starch and fatty alcohol. Renowned for their high tolerance of ionic strength, alkylpolyglycosides are the only class of nonionics that will provide wetting, interfacial tension reduction, and emulsification properties to these systems absent the presence of other classes of surfactants or hydrotroping agents. This property makes them especially valuable when considering how to couple together today's multicomponent pesticide products with disparate properties as where one component is water-soluble and others are oil soluble. The utility is further heightened when it is remembered that both water soluble and oil soluble active ingredients can have their efficacy potentiated by nonionic surfactants. Complex combinations of adjuvant chemistry have also recently entered the marketplace which require tolerance of high ionic strength.

Foam is a property exhibited by most nonionic surfactants, alkylpolyglycosides included. A limited number of agricultural applications benefit from foam but for most formulators using these products the challenge is to control the foam whatever source it comes from. A number of chemical approaches have been investigated and those persons familiar with the chemistry of nonionic surfactants will recognize the similarity to strategies employed with traditional nonionics based on alcohol or alkylarylphenol ethoxylates. The alkypolyglycoside can be based on a low molecular weight fatty alcohol, the degree of polymerization can be extended or truncated and the hydroxyl groups can be functionalized. None of these modifications are without impact on the same properties that attracted us to these surfactants initially and the impact is not always positive.

The focus of the present work is not on structural modification. It is to compare the effectiveness of a range of coformulants and foam control agents available to the

---

[1]Senior Development Chemist, Agrosolutions Group, Cognis Corporation, 4900 Este Avenue, Cincinnati, OH, 45232.

agricultural formulator working with alkylpolyglycoside surfactants. Foam control agents based on fluorochemistry, silicone chemistry, hydrocarbons, fatty alcohols and polymeric defoaming surfactant are compared with each other in the alkylpolyglycoside environment. Relative effectiveness of foam control is assessed and recommendations offered to guide formulation development.

Keywords: alkylpolyglycoside surfactant, alkyl glycoside, foam, foam control, nonionic surfactant, silicone, mineral oil defoamer, organic defoamer, perfluorinated acids, soap, alkoxylated alcohol, linear alcohol, copolymer modified polyols and polymeric defoaming surfactant

Introduction

Alkylpolyglycoside surfactant features are many. They are derived from starch and fatty alcohol renewable resources. They are versatile, mild, environmentally friendly, and their high tolerance of ionic strength is a very important feature that gives the formulator a high degree of latitude in their use. They have high tolerance of ionic strength because solubilization depends on hydroxyl groups instead of ether groups. Alkylpolyglycosides can provide wetting, emulsification properties, efficacy potentiation of actives, and couple in difficult chemistries without the addition of other surfactants and hydrotropes.

Commercial alkylglycosides (Table 1) are available in alkyl chain lengths from the short 8 to 10 carbon chain at an average 1.5 or 1.7 DP (degree of polymerization with glucose) to a midrange 9 to 11 carbon chain at 1.6 DP to a longer 12 to 16 carbon chain at 1.4 and 1.6 DP. There are short carbon chain (C8-10) glycosides provided by other companies. The C9/C11 alkylglycoside designated as AG 91 (AG representing alkylglycoside and 91 representing C9 through C11 carbon chains) is the major focus of this paper. The other alkylglycosides will be designated similarly as AG 80 for C8 to C10, AG 26 for C12 to C16 and AG 86 for C8 to C16.

Foam is a property exhibited by most nonionic surfactants, alkylpolyglycosides included. The formulator is challenged to control foam whatever the source. Controlling foam is an important issue starting at the manufacturing process of the product, to the filling of the product, right up to its use in the final application in the spray tank. Overflowing due to high foam levels could become a safety issue, especially when pesticide actives are present. Foam control approaches are twofold. First, it can be accomplished by structure modification such as using traditional nonionic strategy. One may adjust the hydrophobic or the hydrophilic part of the molecule. This modifies the HLB and cloud point in the typical nonionic. Alkylglycosides can be derivatized but such structure modification often nullifies the desired features of the starting molecule. The second approach to foam control is the addition of foam control agents/coformulants – a simple blending.

The additive approach was adopted in this work. The focus here is to compare the effectiveness of a range of coformulants/foam control agents. The objective is to develop leads for future direction and find materials that are compatible under adverse conditions.

Defoaming mechanisms are best understood with reference to foam stabilization forces explained by Gibbs elasticity E and the Marangoni effect. These have been included in previous symposiums and are explained in literature on foaming. It is the function of a defoamer to counter these stabilization forces. The various defoaming mechanisms that counter these stabilization forces are hydrophobic solids, in-situ formed hydrophobic solids, competition for surfactant by a second liquid phase, low surface tension liquids, antagonistic surfactants, facilitation of lamella draining and disruption of close packing at the air-liquid interface. Some of these categories may overlap or be subsets of each other.

Table 1 - *Commercial alkylpolyglycosides*

Alkyl Chain	**Average DP**	**Key for Charts (Add DP)**
8/10 (45:55)	1.5	AG 80
8/10 (45:55)	1.7	AG 80
9/10/11 (20:40:40)	1.6	AG 91
12/14/16 (68:26:6)	1.4	AG 26
12/14/16 (68:26:6)	1.6	AG 26
8/10/12/14/16 (30:27:22:9:2)	1.6	AG 86

Experimental Details

A low shear foam test was chosen. After comparing various foam tests available we decided on a modified Goldschmidt method. This worked effectively with the many types of defoamers evaluated. It was a quick and reproducible method. This method affords rapid cleaning of test apparatus.

Procedure for Preparation of Sample to Test the Effectiveness of Potential Defoamers in Alkylpolyglycosides

Scope - Since alkylpolyglycosides are available in several active levels (50, 60 and 70%) it will be necessary to reduce them to the same activity level for comparison of effectiveness of a range of potential foam control agents and coformulants. This reduction to a lower active level will help facilitate mixing of a lowered viscosity alkylpolyglycoside and provide room for the addition of defoamers of activity that can be in the range of 10% to 100%. The modified Goldschmidt procedure uses a 100 mL glass wide-mouth, glass stoppered emulsion cylinder. Traces of silicone defoamers were removed using strong alkalis. 2.0% active defoamer was used for most typical defoamers because trial batches made up with higher levels gave excessive separation in some of the blends. Any level lower than 2.0% would not give significant levels of defoaming for comparison purposes in many of the defoamer candidates. Some of the non-standard defoamer materials will be used at up to a 20% active level.

Procedure

1. Calculate the dilution of the alkylglycoside to 40% active by the following formula. This formula will calculate the amount of alkylglycoside in grams to be diluted to 100 grams in the final blend. (100 grams is used here but any multiple or fraction of this amount may be used in order to achieve appropriate mixing volume based on stirrer blade and beaker size chosen.) The required amount of defoamer to achieve the desired active level then the balance as distilled water will be used to dilute the blend to 100 grams.

   Basic principle used in calculations: Grams of Active = Grams of Active
   Solve for x in the following equations. x is the grams of alkylglycoside to dilute to 100 grams.

   If using 50% active alkylglycoside:
   (x) (50%/100) = (100 g)(40%/100)
   x = 80 grams of alkylglycoside (50% active) to dilute to 100 g. (or 40 g to dilute to 50 g or 160 grams to dilute to 200 grams, etc.)

2. Calculate the amount of typical defoamer to give 2.0% active in the 100 g blend of 40% active alkylglycoside by the following formula. Up to 20% active of a non-standard or non-typical defoaming agent may be used. (Again multiples or fractions of this amount may be used. Use the same factor as in #1 above.) Use the table below if % active in defoamer product is an exact multiple of 10, for example 10%, 20% and so on. If not an exact multiple use the following formula to calculate the exact amount.

   Formula:
   Grams (defoamer) = 2.00/(% Activity of defoamer/100)

Table 2 – Use Amount for Concentration of Defoamer

For % Active DF	100	90	80	70	60	50	40	30	20	10
Use Grams per 100 grams	2.00	2.22	2.50	2.86	3.33	4.00	5.00	6.67	10.00	20.00

3. Weigh alkylglycoside into beaker and with a mechanical stirrer such as Heller, Caframo or Heipolh (not a mixer / hotplate) that has the capability to handle the viscosity, mix in calculated amount of defoamer. Then make up the difference to 100 g with distilled water. Be careful not to use too much agitation to create excessive foam that may be hard to dissipate. Note appearance of blend on data sheet as clear, very slightly hazy, slightly hazy, hazy, or cloudy solution.

4. After mixing for approximately 15 minutes immediately use this blend of 40% alkylglycoside active and 2.00% active typical defoamer (higher level of non-typical defoamer agent) to prepare 100 g of 0.10% active alkylglycoside solution. Add 0.25 grams to a beaker and dilute to 100 grams total with distilled water at room temperature. (If this solution sets for any length of time remix with a stirrer and use immediately.) Record on data sheet the appearance of blend as clear or very slightly hazy, slightly hazy, hazy or cloudly dispersion.

5. Place this solution in 25°C environment for at least 15 minutes to achieve 25°C.

6. Test within 2 hours of preparation of this 0.1% active alkylglycoside solution. Remix with hotplate / stirrer immediately prior to using.

7. Test foam performance by adding 25 mL of this solution to a glass stoppered wide mouth 100 ml emulsion cylinder. (Be sure it has been caustic washed if used for previous silicone defoamer testing.)

8. Shake the cylinder vigorously by hand in an up and down motion for 10 seconds.

9. The foam volume is recorded in mls at 15 seconds (enough time to allow cylinder wall drainage), 1 minute, 2 minutes and 5 minutes.

The coformulant/defoamer categories represented that will be blended with the alkylglycosides consist of unmodified silicones which may include silica, organo modified silicones, mineral oils which may include silica, organics, perfluorinated acids, soap, copolymer modified polyols, alkoxylated alcohol, linear alcohol and polymeric defoaming surfactant. Most of these materials are available commercially. Some are not EPA approved inerts. Non EPA approved products were used in order to develop leads or future direction.

The conventions for the following charts are that the y-axis represents foam generated in the shake test, 0 to 100 mL. The x-axis represents time, 15 seconds up to 5 minutes of foam life. The control, foam generated by 0.1% alkylglycoside and no defoamer, will be represented by the farthest bar to the left and the trend line at the top of each chart. Water hardness was at 0 ppm unless otherwise indicated. All tests are conducted at 25°C.

Test Results

Alkoxylated Alcohol Foam Test

The foam test for 0.05% alkoxylated alcohol in 0.1% AG 91 (Figure 1) shows only 30% foam reduction at 5 minutes. A patent on reducing foam in aqueous alkylglycoside compositions (Cole et al. 1996a, Cole et al. 1996b) describes this propoxylated, ethoxylated C12-15 linear alcohol. This coformulant would perform better at higher temperatures above its cloud point of 37° C and take advantage of in-situ formed hydrophobic solids. This temperature is impractical for agricultural applications where well water temperature in northern climates in the spring may be extremely cold.

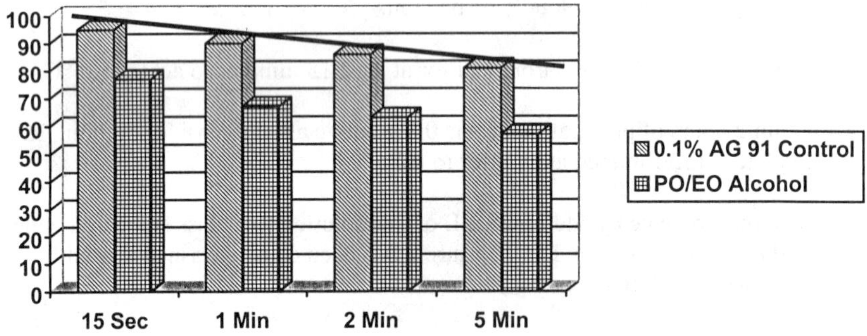

Figure 1 - *Alkoxylated alcohol foaming test at 0 ppm. 0.05% alkoxylate in 0.1% AG 91 at 25 °C. (Milliliters of Foam vs Time).*

Linear Fatty Alcohol Foam Test

The foam test for 0.007% C8 linear fatty alcohol in 0.1% AG 91 (Figure 2) actually shows a slight enhancement of foam. This is the maximum amount of a C8 alcohol that will give a clear solution. Linear fatty alcohols are available commercially in carbon chain lengths of C8 to C18. I understand from a colleague that it is known from his lab experience that low levels of short chained fatty alcohols do enhance foam. This may be due to close packing at the air / liquid interface. C18 alcohol may be more effective but it is less soluble also.

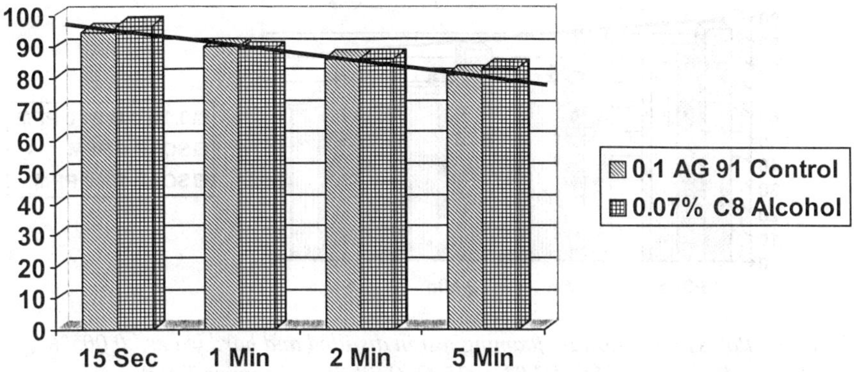

Figure 2 - *Linear fatty alcohol foaming test at 0 ppm. 0.007% C8 alcohol in 0.1% AG 91 at 25 °C. (Milliliters Foam vs Time).*

Potassium Tallowate Foam Test

The foam test for 0.005% potassium tallowate in 0.1% AG 91 (Figure 3) shows some defoaming, about 20% reduction of foam at 5 minutes. Therefore, the test was conducted in hard water to facilitate the formation of insoluble soaps of calcium and magnesium, by nature hydrophobic particles. It was thought they might contribute to the in-situ formed hydrophobic solids defoaming mechanism. There was only a slight improvement from 20% to 30% reduction of foam at 5 minutes. Previous lab bench top experimentation with potassium tallowate in hard water has shown a greater tendency to lower foam in other non-alkylpolyglycoside produced foams. Therefore, it is believed that the alkylpolyglycosides may have emulsified the insoluble soaps formed and inhibited the defoaming mechanism.

Mineral Oil Defoamer Test

The foam test for the mineral oil defoamers, (Figure 4), some containing silica, shows a foam reduction up to 56% with product V but the only product that was clear in a blend with the alkylglycoside was product JMY which did not perform very well.

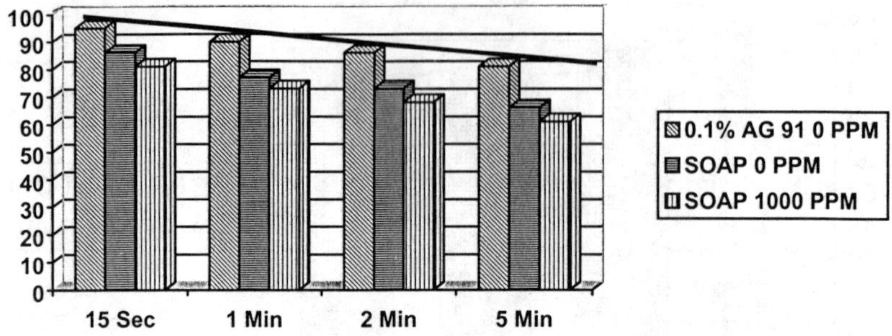

Figure 3 - *Potassium tallowate foaming test in distilled and hard water. 0.005% potassium tallowate in 0.1% AG 91 at 25 °C. (Milliliters Foam vs Time).*

Mineral Oil Defoamer Test

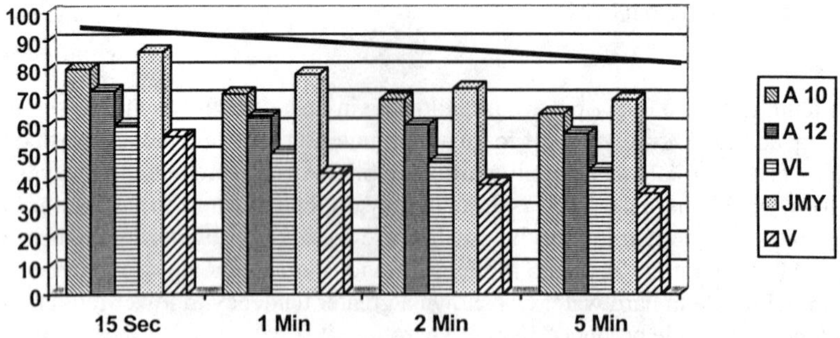

Figure 4 - *Mineral oil defoamer foaming test at 0 PPM. 0.005% mineral oil defoamer in 0.1% AG 91 at 25 °C. (Milliliters Foam vs Time).*

Silicone Oil Defoamer Test

The foam test for the polydimethyl siloxane defoamers (Figure 5), some of which may contain silica, shows up to 98% foam reduction at 5 minutes but again blends with alkylglycoside are not clear, homogeneous, stable systems.

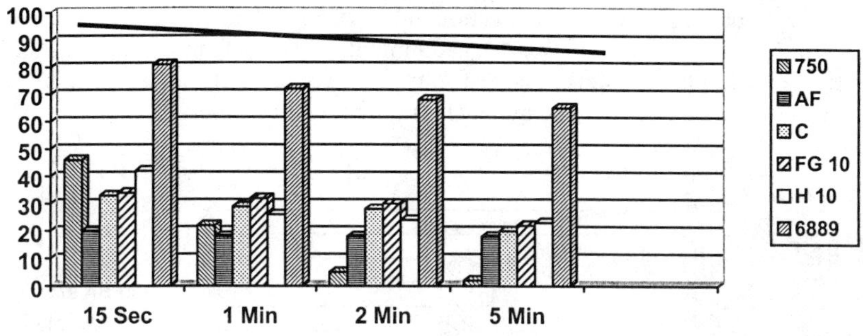

Figure 5 - *Silicone oil defoamer test at 0 PPM. 0.005% defoamer in 0.1% AG 91 at 25 ℃. (Milliliters Foam vs. Time).*

Organo Modified Silicone Test

The foam test for the organo modified silicone defoamers (Figure 6) show up to 75% foam reduction but none are clear aqueous alkylglycoside solutions. Examples here are alkoxylated silicones. These may or may not be polydimethyl siloxane based.

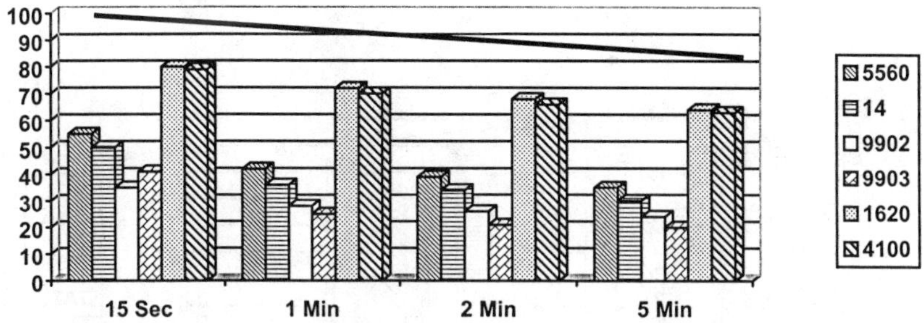

Figure 6 - *Organo modified silicone defoamer test at 0 PPM. 0.005% defoamer in 0.1% AG 91 at 25 ℃. (Milliliters Foam vs. Time).*

Block Copolymer Modified Polyol Test

Block copolymer modified polyols (Figure 7) represent defoamers designed for fermentation systems operating at 30°C. All cloud points are in 20-25°C range. They are designed as insoluble in the media application to be effective defoamers. In this test a maximum of 16% foam reduction was seen at 5 minutes. The defoaming mechanism is believed to be hydrophobic solids. The test solutions were clear. If the temperature was raised to achieve more clouding out, more defoaming might be observed.

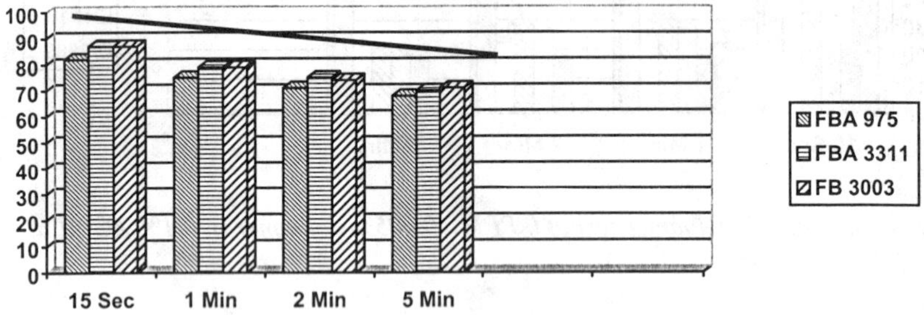

Figure 7--*Block copolymer modified polyols foaming test at 0 PPM. 0.005% defoamer in 0.1% AG 91 at 25 °C. (Milliliters Foam vs. Time).*

Organic Defoamer Test

The foam test for organic defoamers (Figure 8) shows up to 34% foam reduction at 5 minutes but all were cloudy in alkylglycoside aqueous solutions. These products included copolymers, hydrophobic type organics dependent on cloud point for defoaming. They were not polydimethyl siloxane types.

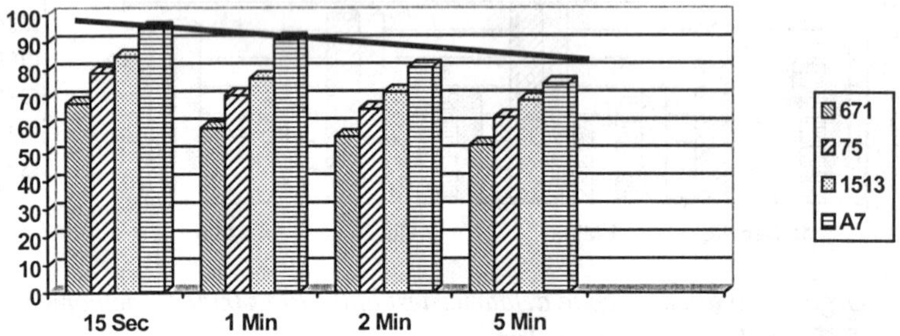

Figure 8 - *Organic defoamers foaming test at 0 PPM. 0.005% defoamer in 0.1% AG 91 at 25 °C. (Milliliters Foam vs. Time).*

Perfluorinated Acids Foam Test

The foam test for perfluorinated acids (Figure 9) shows the performance of two samples of different ratios of perfluorinated phosphinic and phosphonic acids. Both were tested at 0.005% and one at a higher 0.01% level. As opposed to the majority of conventional antifoams which are hydrophobic, the perfluorinated acids are solubilized in the foam film. These compounds work by having extremely low surface tensions of 15-20 dynes/cm in solutions. A presentation at the 20th Symposium (Dookhith, 1999) attributed performance to the lamella draining mechanism. A positive feature of both samples was clarity in alkylglycoside solutions, but the perfluorinated acids did not show the performance desired in this test.

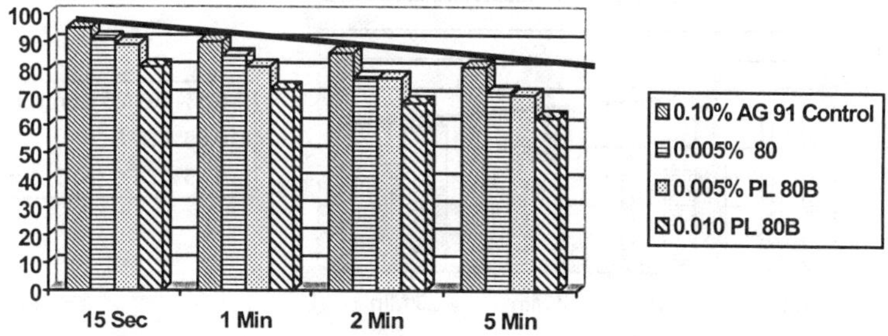

Figure 9 - *Perfluorinated acids foam test at 0 PPM. 0.005% and 0.010% perfluorinated acids in 0.1% AG 91 at 25 ℃. (Milliliters Foam vs. Time).*

Defoaming Surfactant Foam Test

The last category of defoamer tested was a defoaming surfactant designated DFS. 0.005% and 0.05% were tested in 0.1% AG 91 (Figure 10). At the 0.005% level only a very small foam reduction was seen. At 0.05% DFS a significant foam reduction of 73% was seen at 5 minutes. When the sample was aged 50°C for 3 days no reduction in performance was seen. The alkylglycoside solution or blend was clear. This was very encouraging.

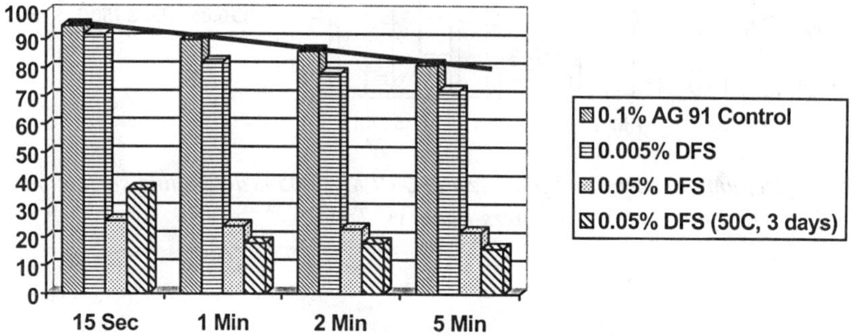

Figure 10 - *Defoaming surfactant foam test at 0 PPM. 0.005% and 0.05% defoaming surfactant (DFS) in 0.1% AG 91 at 25 ℃. (Milliliters Foam vs. Time).*

Testing was continued with the defoaming surfactant DFS in different alkylglycosides. When tested at 0.05% DFS in 0.1% AG 26 with a of DP 1.6 (Figure 11) 32% foam reduction at 5 minutes was observed. The blend was slightly cloudy. The defoaming performance did not decrease when the sample was aged at 50°C for 3 days.

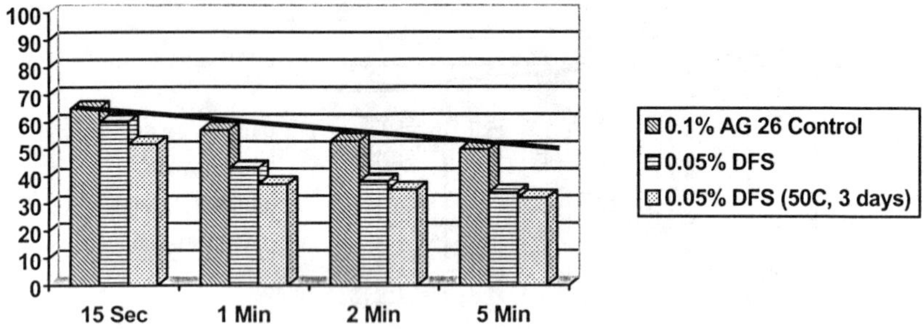

Figure 11 - *Defoaming surfactant foam test at 0 PPM. 0.05% defoaming surfactant in 0.1% AG 26 DP 1.6 at 25 ℃. (Milliliters Foam vs. Time).*

When tested at 0.05% defoaming surfactant DFS in 0.1% AG 26 (Figure 12) 39% foam reduction was seen and the blend was hazy. The defoaming performance decreased slightly for the sample aged at 50°C for 3 days.

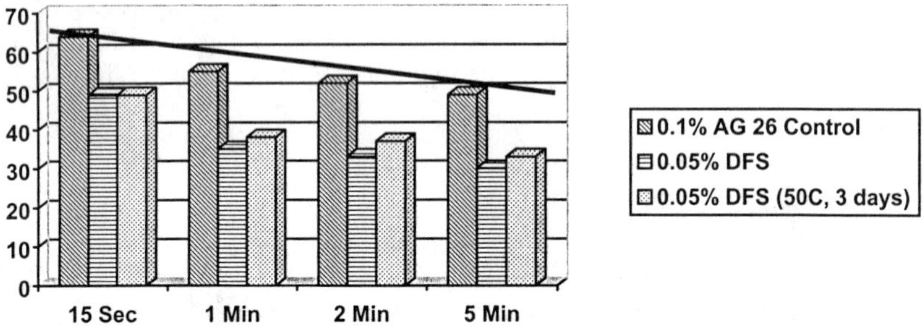

Figure 12 - *Defoaming surfactant foam test at 0 PPM. 0.05% defoaming surfactant in 0.1% AG 26 DP 1.4 at 25 ℃. (Milliliters Foam vs. Time).*

When tested at 0.05% defoaming surfactant DFS in 0.1% AG 80 with DP 1.5 (Figure 13) a significant 81% foam reduction was seen at 5 minutes. The blend or solution was very encouraging since it was clear.

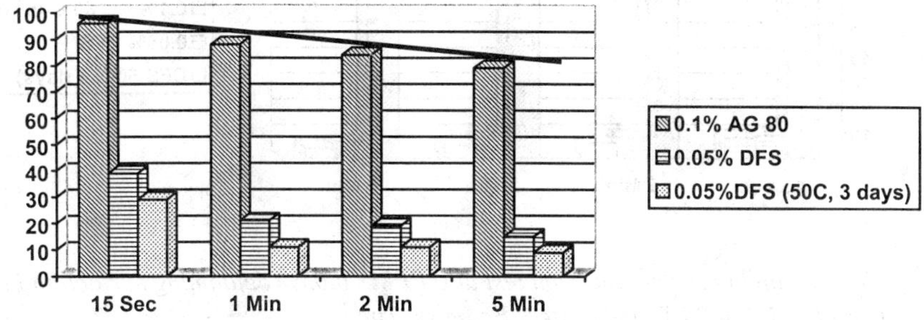

Figure 13 - *Defoaming surfactant foam test at 0 PPM. 0.05% defoaming surfactant in 0.1% AG 80 DP 1.5 at 25 °C. (Milliliters Foam vs. Time).*

When tested at 0.05% defoaming surfactant DFS in 0.1% AG 86 (Figure 14) a 70% foam reduction was seen at 5 minutes and the blend was also clear.

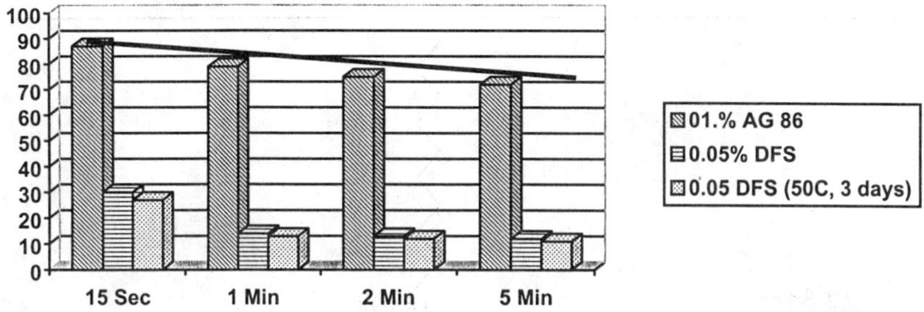

Figure 14 - *Defoaming surfactant foam test at 0 PPM. 0.05% defoaming surfactant in 0.1% AG 86 at 25 °C. (Milliliters Foam vs. Time).*

The last test was conducted on 0.05% DFS in 0.1% AG 80 with DP 1.7 (Figure 15). A 90% foam reduction was seen at 5 minutes. This blend was also clear. In general, aging at 50°C for 3 days showed slight foam reduction improvement in most of the alkylglycosides.

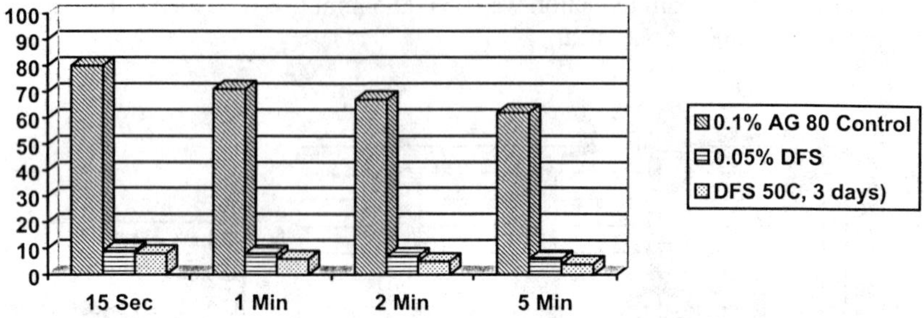

Figure 15 - *Defoaming surfactant foam test at 0 PPM. 0.05% defoaming surfactant in 0.1% AG 80 DP 1.7 at 25 °C. (Milliliters Foam vs. Time).*

The polymeric defoaming surfactant DFS showing this promising performance is shown in a detailed drawing (Figure 16). The defoaming surfactant is novel chemistry (Gross et al. 1998, Gross et al. 1999) that has been TSCA registered and patented. The defoaming mechanism is believed to be disruption of close packing at the air/liquid interface. The shape of the molecule helps one understand why. At the present time this chemistry is not listed as an EPA inert. Still further encouraging news on the surfactant is that it is "well degradable" according to OECD methodology, achieving 70% degradation in 29 days. This is calculated on the basis of chemical oxygen demand.

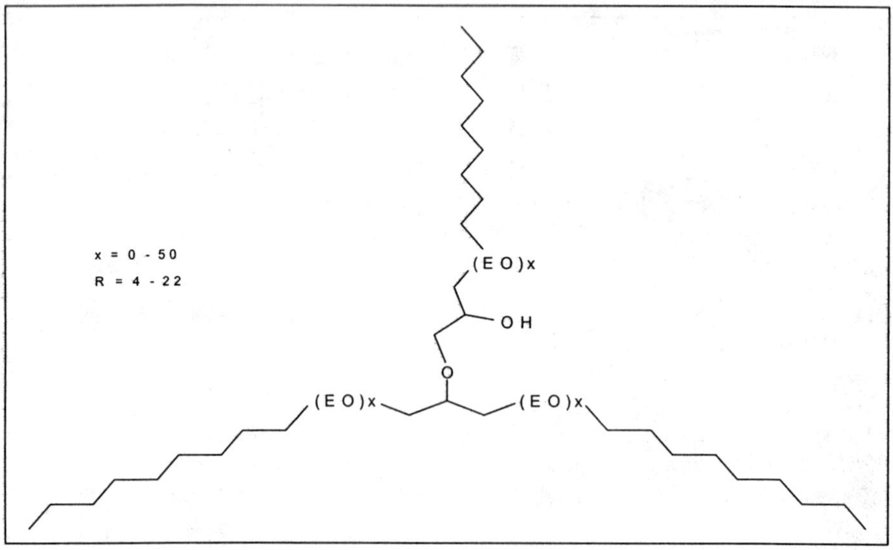

Figure 16 - *Defoaming surfactant (polymeric) (DFS).*

Conclusion

Using this simple technique it was possible to detect the kinds of differences in defoaming we were looking for and compare the effectiveness of a range of coformulants and foam control agents. While much work needs to be done it is believed these interesting leads can be developed into low-foaming, homogeneous alkylglycoside systems.

Acknowledgments

The author sincerely thanks Dr. Roger Garst for his critical review and encouragement. Acknowledgment with appreciation is due to Steve Gross, Tim Morris, Mike Wiggins, Ken Breindel and Benoit Abribat. The author also thanks Angela Abrams for typing the manuscript and Alefesh Hailu for the many lab tests conducted to produce the data.

References

Cole, H., Erler, K. W., Smith, K. L., Francois, B. C., Noble, B.W., and Gross, S. F., 1996a, "Process for Reducing Foam in an Aqueous Alkyl Polyglycoside Composition," U.S. Patent 5,573,707.

Cole, H., Erler, K. W., Smith, K. L., Francois, B. C., and Noble, B.W., 1996b, "Alkyl Polyglycosides in Textile Scour / Bleach Processing," U.S. Patent 5,527,362.

Dookhith, Mohammad T., 1999, "Novel High Performance Water-Soluble Antifoam Based on Perfluoroalkylphosphinic and Perfluoroalkylphosphonic Acids in Crop Protection," *Pesticide Formulations and Application Systems; 20th Volume ASTM STP 1400*, American Society for Testing Materials, West Conshohocken, PA.

Gross, S. F., Wiggins, M. S., Broadbent, R. W., and Devore, D. I., 1998, "Defoaming Compositions," U.S. Patent 5,827,453.

Gross, S. F., Wiggins, M. S., Broadbent, R. W., and Devore, D. I., 1999, "Defoaming Compositions," U.S. Patent 5,895,605.

Frank D. J. Hartmann[1] and Jerry M. Green[2]

Low Dose Urea-Surfactant Adjuvant for Sulfonylurea Herbicides

Reference: Hartmann, F. D. J. and Green, J. M., **"Low Dose Urea-Surfactant Adjuvant for Sulfonylurea Herbicides,"** *Pesticide Formulations and Application Systems: A New Century for Agricultural Formulations, Twenty First Volume, ASTM STP 1414,* J. C. Mueninghoff, A. K. Viets, and R. A. Downer, Eds., American Society for Testing Materials, West Conshohocken, PA, 2001.

Abstract: Urea adducts (urea clathrates) of high molecular weight ethoxylated linear fatty alcohols at low doses are very effective enhancers of sulfonylurea herbicide biological activity. Greenhouse and field trials have identified the preferred alkyl chain length and the degree of ethoxylation for maximum biological efficacy with a broad range of sulfonylurea herbicides at low adjuvant doses. Based on experimental data, the optimum fatty alcohol surfactants include a C_{16-18} hydrophobe with 25 moles of ethoxylation or a C_{24} hydrophobe with 13 moles of ethoxylation. However, the latter alcohol ethoxylate is not readily soluble in water while the former dissolves slowly and forms gel phases when mixed into cold water. These characteristics make them generally unsuitable for use as unformulated materials. One readily cold water soluble and commercially available form of this surfactant type which presents very similar performance is based on a C_{18} linear alcohol ethoxylate with 20 moles of ethoxylation, BRIJ® 78. This ethoxylated linear fatty alcohol remains effective after admixture with related nonionic surfactants and transformation into a readily soluble solid form identified as the urea clathrate of the surfactant.

Urea clathrates can contain between 40-60% of nonionic surfactant by weight while remaining free flowing with melting points above 80 °C. As a result, they can be used as tank-mix or combined with the dry sulfonylurea herbicide formulation. Because of the high activity at low doses, a concentrated dry sulfonylurea herbicide formulation can be achieved with a specific urea clathrate ATPLUS® UCL 1007. Such a combination potentially offers the user more convenience, lower chemical volumes, and reduced costs.

Keywords: adjuvant, clathrate urea, surfactant, sulfonylurea, urea, water dispersible granules

---

[1]Development associate, Uniqema, 900 Uniqema Boulevard, New Castle, DE, 19720.
[2]Biological program leader for Adjuvants, DuPont Crop Protection, Stine-Haskell Research Center, Newark, DE 19714-0030.

Introduction

The current typical adjuvant concentrations in the final spray solution vary depending on the type of adjuvant and the location. In the US, surfactants are usually applied around 0.25% of the spray volume compared to 0.5 - 1% for petroleum and modified seed oil concentrates. Fertilizer adjuvants are commonly used at 2 - 4% of the spray volume.

Surfactant structure-activity analysis showed that the most active surfactants at 0.04% of the spray volume with nicosulfuron on giant foxtail (*Setaria faberi*) had a $C_{16\text{-}18}$ hydrophobe and 12 to 30 moles of ethylene oxide or a C_{24} hydrophobe with 13 moles of ethylene oxide (Green 1999). The most active surfactants identified from this structure activity study were compared versus commercial standards in an adjuvant dose response analysis. The surfactants with a $C_{16\text{-}18}$ hydrophobe with 25 moles of ethylene oxide (cetereath-25) and a C_{24} hydrophobe with 13 moles of ethylene oxide had 50% control doses 0.009% and 0.01% and were 5- to 20-fold more active than currently used commercial adjuvants.

One of the major trends in the US pesticide industry is a shift towards " dry formulations" from conventional "liquid" formulations for various reasons, including flammability, container disposal, and worker exposure. It would be advantageous to use "dry" spray adjuvants with these crop protection agents to coincide with some of these benefits (Underwood et al. 1995). Liquid and soft waxy adjuvants are difficult to include in dry agrochemical formulations, especially when a large amount of adjuvant is required to obtain adequate biological control. Additionally, waxes or solid paste-like adjuvants cannot be used in their current physical form as tank-mix adjuvant. Those waxy solids at room temperature are not readily soluble. They often form gel particles upon mixing in cold water and result in a non-uniform distribution of adjuvant within the spray tank or cause nozzle blocking during application.

Attempts to make dry adjuvants with liquid or waxy adjuvant have used inerts such as clay, silica, or other solid carriers. However, these solid carriers are not water-soluble and may clog fine spray lines and nozzles. Urea forms a solid adduct with a wide range of liquid organic molecules whereby these molecules are being retained in closed cavities or cages provided by the crystalline structure of the urea. Many surfactant molecules have the right size and conformation to form adducts with urea, liquid and "paste-like" ethoxylated alcohols can be transformed into a free-flowing, solid urea complex (Davis 1995).

This study evaluates the possibility to form a urea adduct of highly potent, waxy solid ethoxylated alcohol surfactants and compares their potency with currently used commercial adjuvants.

Materials and Methods

Urea Clathrate Development

This study tested the possibility of incorporating a wide range of alcohol ethoxylates in a urea clathrate (Table 1). The urea clathrates are formed by dissolving urea in the surfactant melt and spray-chilling the melt into small particles resulting in a free flowing powder.

Table 1 - *Alcohol ethoxylate surfactants used in the urea clathrate development.*

Chemical name	HLB
POE-(6)-Synthetic primary C_9-C_{11} alcohol	12.5
POE-(10)- MonoBranched C_{13} alcohol	14.1
POE-(12) -Tridecyl alcohol	14.5
POE-(15)- MonoBranched C_{13} alcohol	15.2
POE-(17)-Cetyl/Oleyl alcohol	14.9
POE-(20)-Cetyl/Oleyl alcohol	15.3
POE-(20)-Stearyl alcohol	15.3
POE-(20)- MonoBranched C_{13} alcohol	16.2
POE-(30)-Cetyl/Oleyl alcohol	16.6

DSC (differential scanning colorimetry) was utilized to assess the presence of free surfactants. This method measures the heat flow (W/g) generated while slowly heating a sample of the material and can therefore detect phase changes within the material. DSC scans were run between -20 °C and 250 °C using a gradient of 4 or 10 °C/min.

General Procedure for Greenhouse Trials

For both experiments weed seeds were planted approximately 2 cm deep in 15 cm plastic pots and thinned after emergence to a uniform stand about 40 seedlings. Plots contained a synthetic growth medium[3] and were watered and fertilized for rapid growth. Artificial metal halide lights with 160 $\mu E/m^2/s$ photosynthetically active radiation supplemented natural intensity during a 16-h photoperiod when light intensity was below 500 $\mu E/m^2/s$. Day temperature was 28± 2 °C and night temperature was 22 ± 2 °C. Relative humidity ranged from 50 to 90%.
Extra pots were planted and used to select uniformity before spraying. Treatments were applied in a 187 l/ha spray volume with a flat fan nozzle[4] at 39 cm with spray pressure set at 275 kPa. Plants were weighed about 14 days after treatment and fresh

[3]Metromix® 350. Grace Sierra Horticultural Products Co., 1001 Yosemite Drive, Milpitas, CA 95035.
[4] TeeJet® flat-fan SS8001E model Spraying Systems Co., Wheaton, IL 60188.

weight reduction was converted to percent control. The studies used the current commercial water dispersible granule formulations of all sulfonylurea herbicides (Table 2).

Table 2 - *Sulfonylurea herbicides used in studies.*

Active ingredient	**Commercial name**
nicosulfuron	ACCENT®[5]
rimsulfuron	TITUS®[5]
metsulfuron methyl	ALLY®[5]
thifensulfuron methyl	HARMONY®[5]
tribenuron methyl	EXPRESS®[5]
chlorimuron ethyl	CLASSIC®[5]

Adjuvant Dose-Response Analysis

The experimental design was a randomized treatment repeated five times. A blend of ethoxylated alcohols[6] and the urea clathrate of this nonionic surfactant blend[7] were compared to a nonionic surfactant commonly used in the US[8] and a nonionic surfactant commonly used in Europe[9]. Adjuvant dose was analyzed with a logistic dose-response equation (Streibig et al., 1993). The 3.1 g/ha nicosulfuron rate was constant. The dose in the model is the adjuvant concentration that varied the biological effect from nearly 0% to nearly 100% control of giant foxtail. The equation relating the biological response "Y" to the adjuvant dose x was

$$Y = C + (D - C) / [exp\ (b\ log\ x - log\ ED_{50})]$$

Were C is the lower limit of control, D the upper limit of control, b the slope, and ED_{50} is the adjuvant dose that gives 50 % control.

[5]ACCENT®, TITUS®, ALLY®, HARMONY®, EXPRESS®, GLEAN®, CLASSIC®, DuPont Crop Protection, Wilmington, DE 19898.

[6]ATPLUS® 6804, ICI Americas, Bridgewater, NJ 08807.

[7]ATPLUS® UCL 1007, ICI Americas, Bridgewater, NJ 08807.

[8] INDUCE® nonionic wetter/spreader adjuvant that contains 90% nonionic surfactant (alkylaryl and alcohol ethoxylate surfactants) and fatty acids and 10 % water. Helena Company, Memphis, TN.

[9]TREND® 90 nonionic surfactant is 90 % isodecyl ethoxylate (7 moles of EO). DuPont de Nemours (France) S.A., Paris, France

Adjuvant Comparison with Various Herbicides and Weeds

The experimental design was a randomized treatment replicated six times. A general linear model procedure in SAS®[10] was used for the analysis. Treatment means for each weed and herbicide grouping were separated by the Waller-Duncan multiple comparison procedure. Treatments with the same letter classification are not significantly different at the 0.05 level.

Results and Discussion

Urea Clathrate Development

A diverse range of alcohol ethoxylates was successfully transformed in a urea clathrate having between 30 - 50 % (w/w) activity resulting in a free flowing powder (Table 3). A typical DSC scan of a urea shows a major narrow peak at 135 °C and a second broad peak above 175 °C (Figure 1). A DSC scan of a urea clathrate has additional peaks around 100 °C and 120 °C (Figure 2). When free surfactant is present, an additional peak will be observed near the melting point of the surfactant. To avoid the presence of free surfactant, the spray conditions and the concentration of the alcohol ethoxylate(s) in the urea clathrate was adjusted in function of the DSC profile.

Table 3 - *Urea clathrates of ethoxylated alcohols.*

Product name	Adjuvant	Concentration (% w/w)
SCS[11] 4707	POE-(6)-Synthetic primary C_9-C_{11} alcohol	30
Atplus UCL 1006	POE-(10)- MonoBranched C_{13} alcohol	50
Atplus S-10	POE-(12) -Tridecyl alcohol	50
Atplus UCL 1003	POE-(15)- MonoBranched C_{13} alcohol	50
Atplus UCL 1004	POE-(17)-Cetyl/Oleyl alcohol	50
SCS 4822	POE-(20)-Cetyl/Ole yl alcohol	50
Atplus UCL 1005	POE-(20)-Stearyl alcohol	50
SCS 5043	POE-(20)- MonoBranched C_{13} alcohol	50
SCS 4823	POE-(30)-Cetyl/Oleyl alcohol	50
Atplus UCL 1007	Atplus 6804	50

The activity of SCS 4707 was lowered to 30% to ensure a melting point of the urea clathrate above 80 °C. It was not possible to form a urea adduct of ethoxylated alcohol's having a lower degree of ethoxylation.

---

[10] SAS® institute, Inc., Box, Cary, NC 27511
[11] SCS : Development product code Uniqema Europe

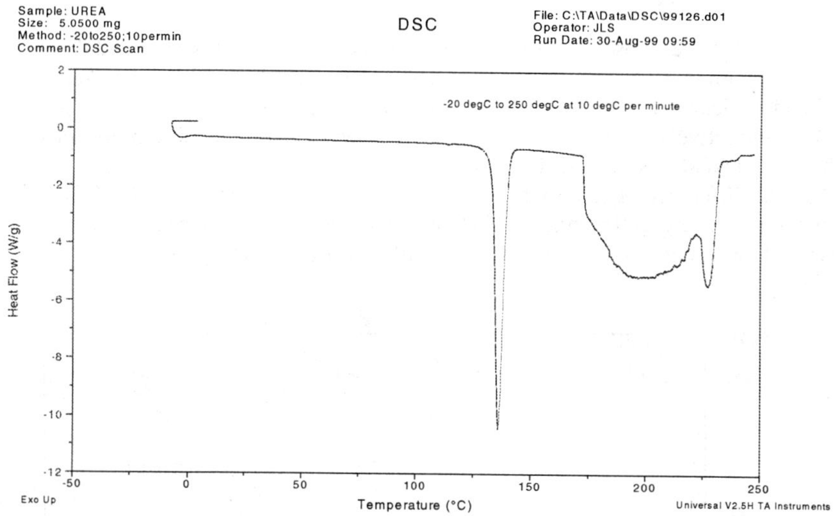

Figure 1 - *DSC-scan of urea.*

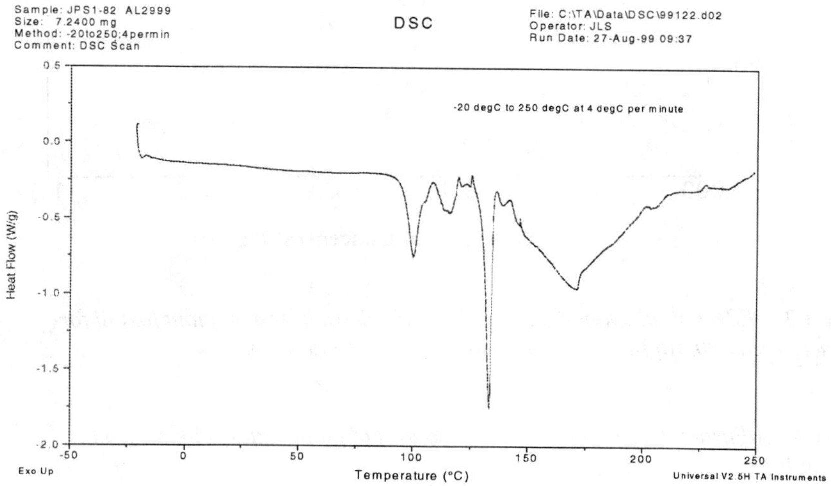

Figure 2 - *DSC-scan of Atplus UCL 1007.*

Adjuvant Dose-Response Analysis

The dose-response analysis compared the activity of Atplus 6804 and Atplus UCL 1007 with two commercial standard surfactants, Trend 90 and Induce. Atplus 6804 and Atplus UCL 1007 had 50% control doses of 0.016 % and 0.036% of spray volume, respectively (Table 4). Atplus UCL 1007 was almost four-fold more active than Induce and almost three-fold more active than Trend 90.

The difference in activity between Atplus 6804 and Atplus UCL 1007 is only related to the difference in actual surfactant concentration between both products. The activity of the adjuvants tested increased similarly with increasing dose to equivalent maximum levels (Figure 3).

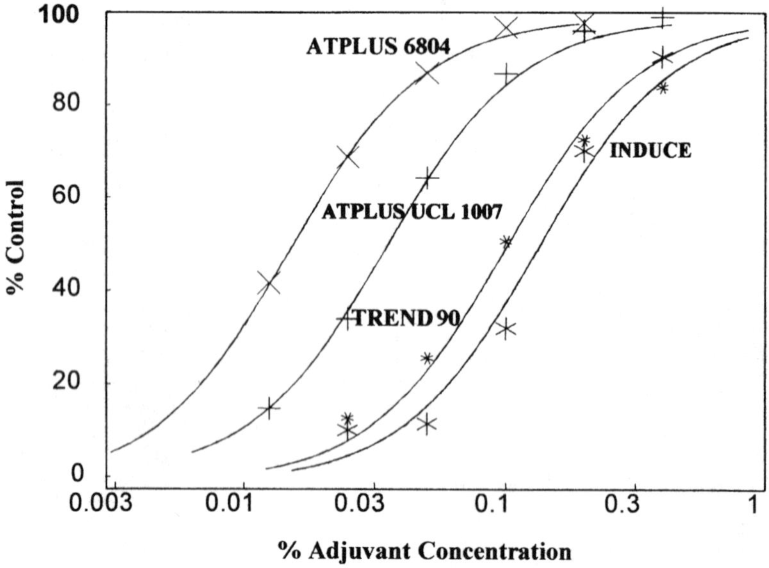

Figure 3 - *Effect of adjuvant dose with 3.1 g/ha nicosulfuron on giant foxtail for Induce, Trend 90, Atplus UCL 1007, and Atplus 6804 adjuvants.*

Table 4 - *Adjuvant dose needed for 50 % control of giant foxtail with 3.1g/ha nicosulfuron.*

Adjuvant	50 % control dose (% spray volume)	95 % confidence interval (% spray volume)
Atplus UCL 1007	0.036	0.032 - 0.040
Atplus 6804	0.016	0.014 - 0.017
Trend 90	0.1	0.08 - 0.12
Induce	0.14	0.12 - 0.16

Figure 4 - *Effect of Atplus UCL 1007 versus Atplus 6804.*

Figure 5 - *Effect of Atplus UCL 1007 versus Induce.*

Adjuvant Comparison with Different Sulfonylurea Herbicides and Weeds

Atplus UCL 1007 and Trend 90 were compared at 0.1% of the spray volume with a range of sulfonylurea herbicides on several weed species (Table 5). In all cases, Atplus UCL 1007 significantly enhanced the activity more than the herbicide alone and gave equal or better control than the Trend 90 treatment even though it only contained about half the amount of surfactant. On barnyardgrass (*Echinochloa crus-galli*) Atplus UCL 1007 gave significantly more activity with nicosulfuron and rimsulfuron. On common lambsquarters (*Chenopodium album*), both adjuvants gave 95% or more control with low rates of metsulfuron methyl, thifensulfuron methyl, and tribenuron methyl. Atplus UCL 1007 significantly enhanced nicosulfuron activity more than Trend 90 and enhanced thifensulfuron and chlorimuron statistically similarly on velvetleaf (*Abutilon theophrasti*).

Table 5 - *Effect of Atplus UCL 1007 and Trend 90 on three species with a range of sulfonylurea herbicides.*

Weed Species	Herbicide treatment	Adjuvant Treatment	% Control	Statistical T groupings
Barnyardgrass	2 g/ha rimsulfuron	None	0	a
		0.1% w/w Trend 90	41	b
		0.1% w/w Atplus UCL 1007	67	c
	4 g/ha nicosulfuron	None	0	a
		0.1% w/w Trend 90	19	b
		0.1% w/w Atplus UCL 1007	51	c
Common lambsquarters	1 g/ha metsulfuron	None	12	a
		0.1% w/w Trend 90	96	b
		0.1% w/w Atplus UCL 1007	95	b
	2 g/ha thifensulfuron	None	34	a
		0.1% w/w Trend 90	94	b
		0.1% w/w Atplus UCL 1007	95	b
	2 g/ha tribenuron	None	51	a
		0.1% w/w Trend 90	97	b
		0.1% w/w Atplus UCL 1007	95	b
Velvetleaf	1 g/ha thifensulfuron	None	62	a
		0.1% w/w Trend 90	83	b
		0.1% w/w Atplus UCL 1007	88	b
	2 g/ha chlorimuron	None	58	a
		0.1% w/w Trend 90	72	b
		0.1% w/w Atplus UCL 1007	79	b
	16 g/ha nicosulfuron	None	13	a
		0.1% w/w Trend 90	57	b
		0.1% w/w Atplus UCL 1007	79	c

Conclusions

Optimizing the adjuvant surfactant for sulfonylurea herbicides shows that a large reduction of application dose is possible compared to currently used commercial liquid tank-mix adjuvants.

Transforming these waxy to "paste-like" surfactant blends into urea clathrates does not change their biological activity with sulfonylurea herbicides which is only related to the difference in actual surfactant concentration between products.

The formation of highly potent, low dose urea-surfactants adjuvants provides a high quality dry formulation that could be used as a tank-mix adjuvant or put in a water dispersible granule herbicide formulation.

References

Davis, R. 1995, Solid adjuvants based on urea-surfactant adducts, "*Pesticides Formulations and Applications Systems*" 15th volume, ASTM STP1268, Collins H. M., Hall, F.R. and Hopkinson M. eds., ASTM, West Conshohocken, PA, pp.161-167.

Green, J. M.. 1999, Optimizing alcohol ethoxylate surfactant activity at low doses, *Weed Technology*, Volume 7, pp 737-740.

Streibig, J.C., Rudemo M. and Jensen J. E.. 1993, "Dose-response curves and statistical models." In J.C. Streibig and P. Kudsk. Eds, *Herbicide Bioassays*, CRC Press, Boca Raton, FL, pp. 30-55.

Underwood A., A. Clark, R. E. Mack, J. Thomas, J. R. Roberts, G. Volgas. 1995, Dry Concentrate (DC) spray adjuvants, *Fourth International Symposium on Adjuvants for Agrochemicals*, Melbourne, Australia, (FRI Bulletin No 193). pp. 391-396.

Richard F. Burow,[1] Donald Penner,[2] Frank C. Roggenbuck,[2] and David E. Wallick[3]

A New Adjuvant Blend for Glyphosate Herbicides

Reference: Burow, R. F., Penner, D., Roggenbuck, F. C., and Wallick, D. E., **"A New Adjuvant Blend for Glyphosate Herbicides,"** *Pesticide Formulations and Application Systems: Twenty First Volume, ASTM STP 1414,* J. C. Mueninghoff, A. K. Viets, and R. A. Downer, Eds., American Society for Testing and Materials, West Conshohocken, PA, 2001.

Abstract: With the advent of glyphosate-resistant crop seed, glyphosate has become the herbicide of choice for controlling weeds in a wide range of field crops. Many manufacturers of agricultural chemicals are adding glyphosate herbicides to their herbicide product lines to serve this market need. An adjuvant blend has been developed specifically to enhance the performance of these glyphosate products.

Organosilicone (OS) surfactants are effective adjuvants to boost the performance of a number of foliar-applied herbicides. They are effective adjuvants when used with glyphosate herbicides against a number of weed species, but are weak or antagonistic on certain weeds, particularly lambsquarters and certain perennial grasses. Alkyl diphenyl oxide disulfonate (ADODS) surfactants, on the other hand, are effective on many of the weeds that are not enhanced by the organosilicone surfactants. It has been found that combinations of the organosilicone and ADODS surfactants are effective across a broad range of weed species.

In greenhouse studies using a low rate of glyphosate, the use of a combination of a mixture of OS and ADODS surfactants provided control of a broad range of weed species and provided rainfastness as well.

Keywords: organosilicone, adjuvant, trisiloxane, alkyldiphenyloxidedisulfonate, surfactant, glyphosate, herbicide

Introduction

With the introduction of glyphosate-resistant crops, glyphosate has become more popular as a herbicide for weed control in a wide range of field crops. Many manufacturers of agricultural chemicals are adding glyphosate herbicides to their herbicide product lines to serve this market need. Many of these glyphosate products

---

[1]Associate Development Scientist, Dow Corning Corporation, Midland, MI 48686-0994.

[2]Professor and Research Technician, respectively, Crop and Soil Sciences, Michigan State University, E. Lansing MI 48824.

[3]Research Leader, The Dow Chemical Company, 667 Building, Midland, MI 48667.

require an adjuvant, either included in the formulation, or added as a tank-mix additive, to assure good control of the broadest range of weeds.

Organosilicone (OS) surfactants used in agriculture are organosilicone polyether surfactants having the general structure

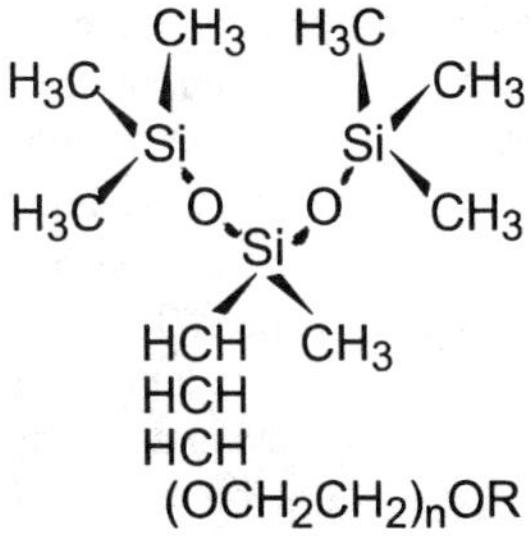

where n = 7 or 8 and R = H, CH3, or COCH3. OS surfactants are effective adjuvants to boost the performance of a number of foliar-applied herbicides (Penner et al. 1999, Burow el al. 1995). When used with glyphosate herbicides, OS surfactants enhance efficacy against a number of weed species, but are weak or antagonistic on certain other weeds, particularly lambsquarters and certain perennial grasses (Gaskin and Stevens 1993).

Alkyl diphenyl oxide disulfonate (ADODS) surfactants are unique disulfonated structures with the general structure shown below.

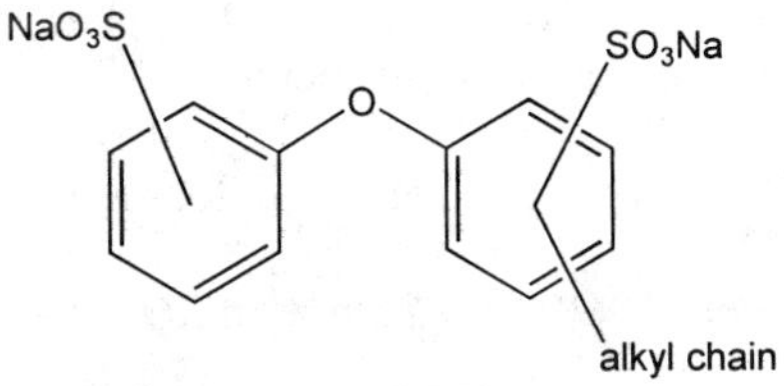

ADODS surfactants exhibit a broad array of surfactant performance characteristics depending on the size of the alkyl hydrophobe side chain. Particular ADODS surfactants have been found to enhance the super-spreading of organosilicone surfactants (Wallick, 1998). The ADODS surfactants have been evaluated as adjuvants for glyphosate (Holoman and Quencer 1992) and found to enhance efficacy on many of the weeds that are not enhanced by organosilicone surfactants.

The purpose of this study was to determine whether or not a combination of OS and ADODS surfactants would enhance the performance of glyphosate herbicides across a range of problem weed species. The specific objectives were to evaluate mixtures of OS and ADODS surfactant for efficacy with glyphosate across three problem weed species and to compare the mixture to the ROUNDUP ULTRA® formulation. (ROUNDUP ULTRA is a registered trademark of Monsanto Co, 800 Lindbergh Blvd., St. Louis, MO 63167.)

Methods and Materials

Evaluation of Surfactant Mixture across Weed Species

Velvetleaf, common lambsquarters, and giant foxtail seeds were obtained from V&J Seed Farm, Woodstock, Illinois and planted in Baccto professional potting mix in 946-ml plastic pots. After the seedlings emerged, the plants were thinned to one plant per pot for velvetleaf, two plants per pot for common lambsquarters, and three plants per pot for giant foxtail. The plants were watered and fertilized as needed. The plants were maintained in the greenhouse at 25±2°C with a 16 hr day/8 hr night with supplemental lighting from sodium vapor lamps. When the velvetleaf and giant foxtail were in the four-leaf stage and the common lambsquarters in the six-leaf stage, they were sprayed with the test treatment solution using a link-belt sprayer delivering 246 L/ha at 175 kp. The herbicide was applied at a low rate to maximize differences between treatments. Glyphosate (ACCORD® -- Registered: Monsanto Co.) plus the nonylphenol ethoxylate X-77 (Loveland Industries) was included for comparison purposes. Ammonium sulfate at 2% in the spray solution was applied with the glyphosate to all adjuvant treatments. All test adjuvants were applied at 0.25% (v/v) of the spray solution, as was X-77. Adjuvants tested were OS surfactant (SYLGARD® 309 Silicone Surfactant – Registered: Dow Corning Corp.) and two different ADODS surfactants (DOWFAX® Surfactants – Registered: Dow Chemical Co.) and 25/75 mixtures of these. The ADODS surfactants were supplied as water solutions, so a greater amount was added to correct for the water content. Glyphosate as ACCORD was applied at 0.11 kg/ha to velvetleaf and giant foxtail and at 0.34 kg/ha to common lambsquarters. The very low rates of glyphosate were used to provide the greatest differentiation between treatments.

All plants were evaluated for visual injury 14 days after herbicide application. Data presented are the means of two experiments with four replications in each.

Comparison of Glyphosate Plus ADODS Surfactants to ROUNDUP ULTRA®

The experimental procedures for this study were the same as for the last with the following changes. The new adjuvants were applied at 0.25% of the spray solution as was X-77 and the OS surfactant. To determine rainfastness, 2.54 cm of simulated rainfall was applied one hour after treatment. Glyphosate as ACCORD was applied to the various weed species at the same rates as above. ROUNDUP ULTRA was included as a control. ROUNDUP READY® corn and soybean received glyphosate at the 1X rate of 1.1 kg/ha a.i.

The weeds were evaluated for visual injury 14 days after herbicide application. The corn and soybeans were evaluated for visual injury up to 11 and 12 days after treatment, respectively. The data presented are the means of two experiments with four replications each.

Results and Discussion

Evaluation of Surfactant Mixtures across Species

As expected from prior experiments, the glyphosate treatment containing the OS surfactant produced significantly greater injury to velvetleaf than the X-77 treatment and all the ADODS and ADODS blend treatments (Figure 1). The treatment using 25% OS surfactant and 75% ADODS A was almost equal to the treatment with OS surfactant alone.

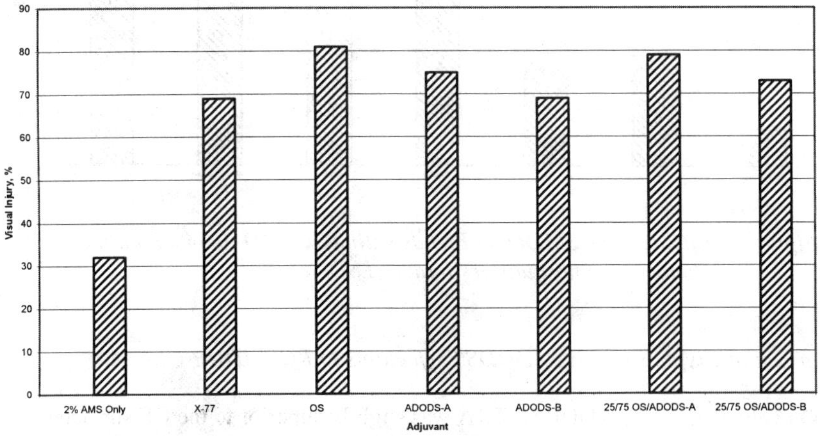

Figure 1. Efficacy of OS/ADODS Blends with ACCORD (0.11 kg a.i./ha) -- Velvetleaf Injury. $LSD_{0.05} = 6$.

As Figure 2 shows, the OS and 25/75 OS/ADODS-A were the most effective treatments on giant foxtail.

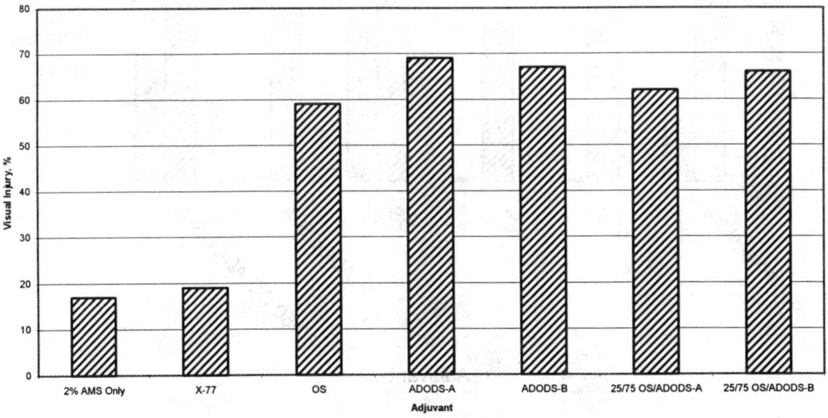

Figure 2. Efficacy of OS/ADODS Blends with ACCORD (0.11 kg a.i./ha) -- Giant Foxtail Injury. $LSD_{0.05} = 12$

On common lambsquarters, the treatment containing the OS surfactant produced essentially the same degree of injury as the X-77 treatment. The two ADODS surfactants produced results superior to the X-77 and the OS surfactant (Figures 3). The OS/ADODS blends were essentially equal to the ADODS only treatments.

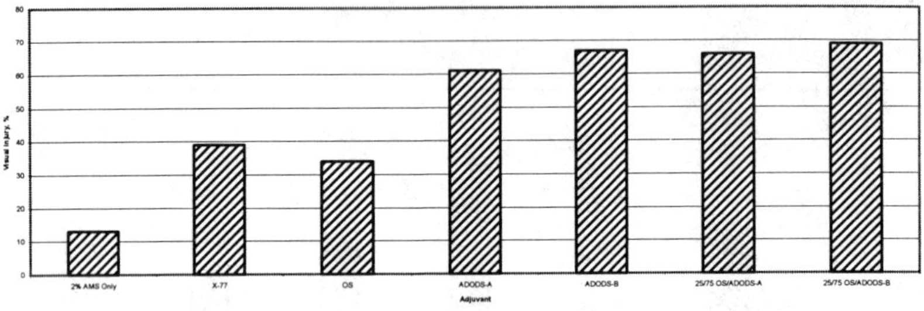

Figure 3. Efficacy of OS/ADODS Blends with ACCORD (0.34 kg a.i./ha) – Common Lambsquarters Injury. $LSD_{0.05}$ = 8

Comparison of Glyphosate Plus ADODS Surfactants to ROUNDUP ULTRA®

On velvetleaf, ROUNDUP ULTRA was slightly superior to the OS surfactant, but several of the ADODS and OS/ADODS blends were equivalent (Figure 4). When 2.54 cm of simulated rainfall was applied 1 hour after treatments, all of the treatments containing OS surfactant were superior to ROUNDUP ULTRA.

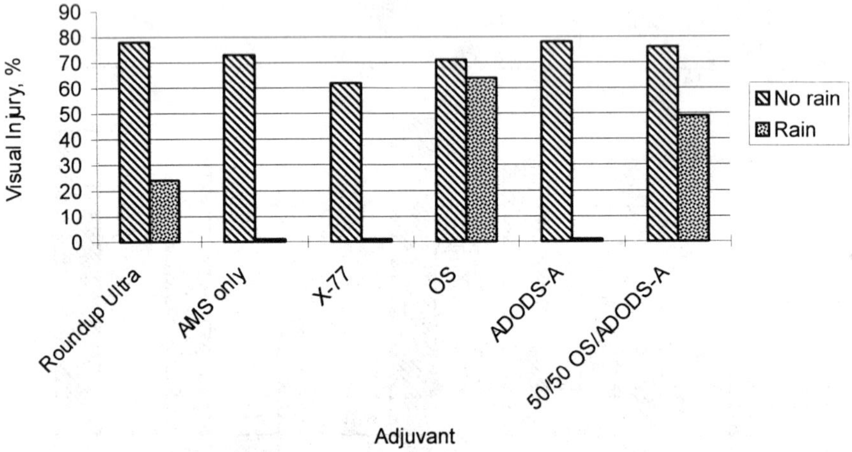

Figure 4. Comparison of OSADODS-A Blends with ACCORD and ROUNDUP ULTRA (0.11 kg a.i./ha) – Velvetleaf injury. $LSD_{0.05}$ = 6.

On giant foxtail, the ADODS surfactants were superior to ROUNDUP ULTRA, and the OS/ADODS blends were as well (Figure 5). All treatments containing OS and/or ADODS surfactants appear to provide rainfastness equal or superior to ROUNDUP ULTRA on giant foxtail.

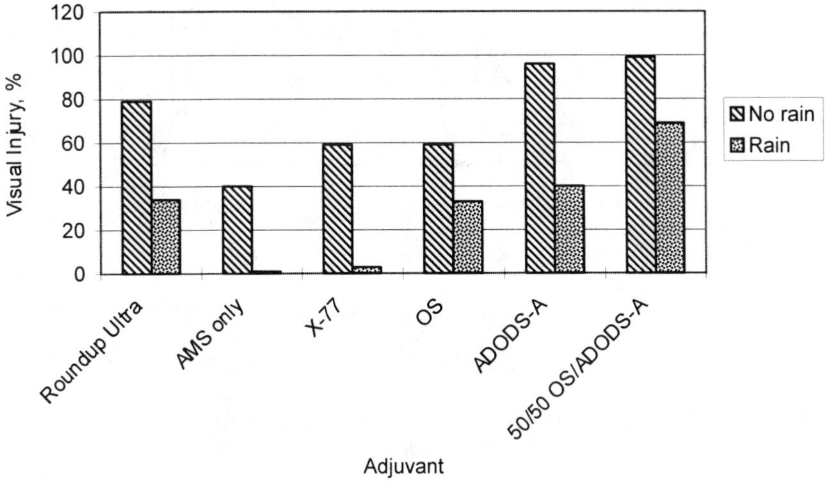

Figure 5. Comparison of OS/ADODS-A Blends with ACCORD vs. ROUNDUP ULTRA (0.11 kg a.i./ha) – Giant Foxtail Injury. $LSD_{0.05} = 12$

On common lambsquarters, the ADODS surfactants provided results superior to the ROUNDUP ULTRA treatment (Figure 6). The OS/ADODS blends were less effective than ROUNDUP ULTRA. Rainfastness with all of the OS and/or ADODS treatments was superior to ROUNDUP ULTRA but was less than desired. Greater differentiation of treatments could probably have been achieved if a higher rate of glyphosate had been used for the common lambsquarters treatments.

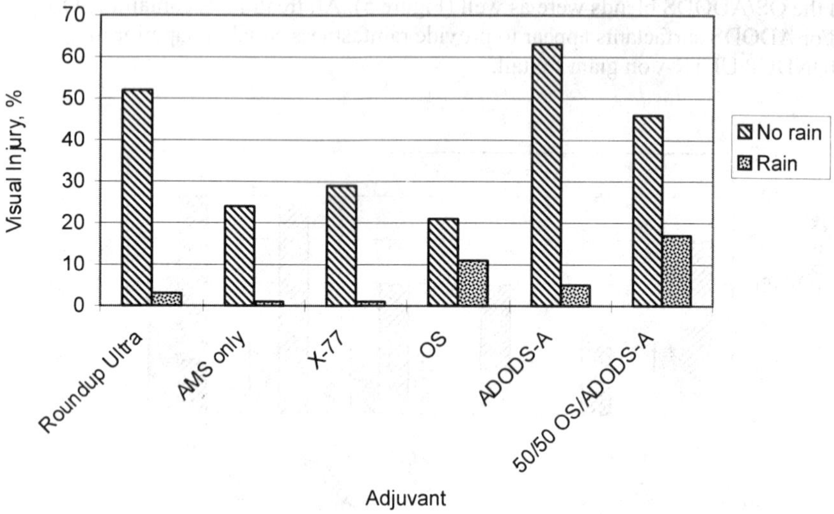

Figure 6. Comparison of OS/ADODS-A Blends with ACCORD vs. ROUNDUP ULTRA (0.34 kg a.i./ha) – Common Lambsquarters Injury. $LSD_{0.05}$=6.

None of the adjuvants applied with glyphosate caused injury to ROUNDUP READY corn or soybeans up to 11 and 12 days after treatment.

Conclusion

The two greenhouse studies demonstrated that a blend of OS surfactant and ADODS surfactant can provide performance equal to, and in some cases, superior to performance of ROUNDUP ULTRA. In most cases, the blend provides much greater rainfastness than ROUNDUP ULTRA. These blends provide the formulator and the user of glyphosate herbicides a means to obtain control of a broader range of weeds and also achieve rainfastness. The same may be true for glufosinate formulations as well.

The positive results obtained in the greenhouse studies led to the decision to make ADODS A surfactant available commercially to the agricultural chemicals market as DOWFAX AG Surfactant.

References

Burow, R. F., Penner, D., Roggenbuck, F. C., and Hill, R. M., 1995, "Relationship of Organosilicone Adjuvant Structure and Phase Behavior to Activity Enhancement of Acifluorfen and Glyphosate," *Proceedings of the Fourth International Symposium on*

Adjuvants for Agrochemicals, Melbourne, Australia (New Zealand Forest Research Institute Bulletin 193), pp 54-59.

Gaskin, R.E., and Stevens, P.J.G., 1993, "Antagonism of the Foliar Uptake of Glyphosate into Grasses by Organosilicone Surfactants. Part 1: Effects of Plant Species, Concentrations, and Timing of Application," *Pesticide Science*, Vol. 38, pp. 185-192.

Holoman, Jr., S., and Quencer, L.B., 1992, "The Use of Dowfax Surfactants to Improve the Efficacy of Herbicide Products," *Adjuvants for Agrichemicals*, C.L. Foy, Ed., CRC Press, New York, pp. 533-543.

Penner, D., Burow, R. F., and Roggenbuck, F. C., 1999, "Use of Organosilicone Surfactants as Agrichemical Adjuvants," *Silicone Surfactants*, Randal M. Hill, Ed., Marcel Dekker, New York, pp. 241-258.

Wallick, D. E., 1998, Unpublished Internal Laboratory Data, Dow Chemical Company, Midland, MI.

Author Index

Subject Index